smokechasing

smoke

Stephen J. Pyne

chasing

The University of Arizona Press

Tucson

The University of Arizona Press

First printing

∞ This book is printed on acid-free, archival-quality paper.

Manufactured in the United States of America

08 07 06 05 04 03   6 5 4 3 2 1

Library of Congress Cataloging-in-Publication Data
Pyne, Stephen J., 1949–
Smokechasing/Stephen J. Pyne.
p. cm.
ISBN 0-8165-2284-7 (cloth: alk. paper)—
ISBN 0-8165-2285-5 (pbk.: alk. paper)
1. Forest fires—United States—Prevention and control.
2. Fire prevention—United States.   3. Fire ecology—United States.
4. Forest fires—Prevention and control.   5. Fire prevention.
6. Fire ecology.   I. Title.
SD421.3.P963   2003
634.9'618–dc21   2002011504

British Library Cataloguing-in-Publication Data
A catalogue record for this book is available from the British Library.

To Sonja, who still remembers the smoke,
Lydia, who saw through it,
and Molly, who would have enjoyed the chase

"Line up, shut up, and go look for smokes!"

—Superintendent to an unruly hot shot crew
on the Demotte fire

*He said that the boy should find that place where acts of
God and those of man are of a piece. Where they cannot be
distinguished.*

Y qué clase de lugar es éste? the boy said.

*Lugares donde el fierro ya está en la tierra, the old man
said. Lugares donde ha quemado el fuego.*

—Cormac McCarthy, *The Crossing*

contents

burning off

back at the cache

paperwork

debriefing

why
smokechasing

an
author's
note

Painting, architecture, politics, even gardening and golf—all have their critics and commentators. Fire does not. Its literature resides in government reports, scientific papers, a recent rash of personal memoirs, and a journalism that typically has more in common with the sports page than with arts and ideas (as though wildland firefighting has become a government-sponsored extreme sport). Fire deserves better. Its intellectual content is at least as gripping as baseball stories and its narratives much more fiery than aging Bauhaus buildings. And that is the rationale behind this book: it is a volume of commentaries by a friendly fire critic.

The genre itself deserves standing. Only recently has anything like a fire culture of books appeared—of histories, imaginative fiction, personal essays, memoirs, critical exegeses, and so on. That outpouring can be dated more or less precisely to the appearance of Norman Maclean's *Young Men and Fire* in 1992. The vast runoff of fire experiences gets channeled mostly into a very few, familiar, deep-scoured gullies—the firefight as battlefield, the conflagration as social disaster (akin to earthquakes, floods, *Titanics*), and more recently fire as an ecological celebrity. This book proposes a more robust literary inquiry. I've tried to analyze fire as I would an art moderne house, an election campaign, or a rereading of *Ulysses*.

A few of these essays have appeared previously in print. Approximately half are altogether new. All have been revised, some nearly beyond recognition. This scattering of pieces has argued for a less-massive structure than my continental-scale histories. Those narratives resemble suspension bridges; a good bit of the prose goes to bear up and align the varied stresses into a coherent whole. A flotilla of smaller pieces, each more or less intact by itself, however, pleads instead for the literary equivalent of a pontoon bridge, joined, in this instance, by the flexible metaphor of smokechasing. The pieces should float separately, yet together help span the grand theme of fire on Earth. The repetitions between them are, I trust, simply the necessary linkages to keep the composite structure together.

This book builds on a previous anthology, *World Fire: The Culture of Fire on Earth*, and likewise has borrowed a master metaphor to help bond its unruly brood. *Smokechasing* is an American term coined to describe the practice of sending one or, later, two firefighters into wildlands to track down the source of a reported smoke. The organization of this book roughly traces the stages of such an event.

Smokechasers have formal safety rules to follow, codified into the Ten Standard Firefighting Orders. For the book, "Standard Orders" refers to some basic premises and definitional guidelines, among them what fire is.

At some point, a smokechaser typically will follow a compass bearing cross-country. The source of the bearing may vary. It may be taken from a ridge or a treetop. If it comes from an aerial observer, it must be "corrected" to accommodate the local variation (or declination) from true north. Here, the section titled "Bearings" offers a global survey of fire topics, supplementing the "hotspotting" essays in *World Fire*.

The next two sections deal specifically with the U.S. scene. An actual firefight, if those involved are lucky, will consist of an initial attack that successfully contains the smoke-sourcing fire while it remains small. If that fails, however, suppression moves into an extended attack. *World Fire* offered an update of U.S. fire history in an essay titled "Initial Attack." The essays grouped here under "Extended Attack" continue and elaborate on that inquiry. All deal with fire control. "Burning Off," however, considers the complementary role of controlled burning. Again, its focus is the United States.

Once a fire is controlled or a prescribed burn completed, crews return to the fire cache, which for wildland firefighters serves the same role as a firehouse for urban firefighters. The cache is the place to regroup, to recondition tools and reoutfit vehicles, to talk over what has happened, to swap lies and humorous stories, and to reflect. "Back at the Cache" thus encompasses a cluster of essays that look a bit reflectively and perhaps whimsically at our species' experience with fire. One piece, from which the book derives its title, uses the metaphor of smokechasing to explore the character of environmental history as practiced in the academy.

Always there is the paperwork. As with any bureaucracy, there are reports to file, narratives to write, forms to complete. In this context, "Paperwork" deals with the creation of something resembling a formal literature on fire and, for good measure, offers a look at movies.

Finally, it is good practice to debrief after a burn. The telling alone has its value, even its catharsis. One can review not only the craft of smokechasing but its policies or, here, its purpose.

That done, one is ready to begin anew. Which is where any bred-in-the-bone smokechaser wishes to be.

As always, there are many persons to thank. The range is especially broad in this instance because so many of the essays date from travels, for which I am beholden to all who made such trips possible and who hosted me. Other essays originated from invitations. But a special nod of appreciation goes to my daughter, Lydia Pyne, who, after hearing me grouse about having so many stray ideas and nowhere to go with them, suggested I gather all the mavericks together into a book and then did a first edit.

Others who assisted in one form or another include H. J. "Doc" Smith, J. G. Goldammer, Neil Burrows, Rick Sneeuwjagt, Scot Slovic, Dante Arturo Rodriguez Trejo, Alfred Siemen, Sergio Guevara, Leonard Brennan, Liam Fogarty, Kristan Cockerill, Tony Burgess, Kwesi Orgle, Victor Agyeman, Alex Asare, Kennedy Owusu-Afriyie, Godfred Ohene-Gyan, Lucy Amissah, Mark Breighton, Jason Greenlee, Peter Warshall, Tom Fenske, Hutch Brown, Jack Dieterich, Stephen Gale, Steven Anderson, "Packer Bill" Wyman, Alianor True, and Annie Barva. My thanks to them all—and not least to my wife, Sonja, who still manages to find meaning in the flames.

smokechasing

standard
orders

the
element
that
isn't

How should we think about fire? An answer is not obvious. It is testimony to
the immense significance of fire that humanity has for so long chosen not only
to anthropomorphize it but to grant it a substantive identity it does not deserve.
Early philosophers considered it a god or at least theophany, the manifestation
of a godlike presence and power. The Aztecs called it Huehueteotl (or Dios
Viejo, the Old God), and the Hindus, Agni, along with Indus the most vener-
able of their pantheon. The ancient Greeks and the ancient Chinese labeled
it an element. For Western civilization, it then morphed into a declination of
lesser substances such as phlogiston and caloric before ending as a subservient
chemical reaction, the rapid oxidation, usually accompanied by flame, of other
substances. Today it no longer claims reality as an autonomous substance.
Rather, fire is a phenomenon that derives from its circumstances. It is what
results when heat, fuel, and oxygen combine under suitable conditions. It is
a reaction, a process. It has no reality apart from the physical circumstances
that make it possible. It synthesizes its surroundings. And that, in brief, is
equally the lesson of its intellectual history. Fire's definition has changed with its
cultural circumstances. It takes its character from its context.

In this way, fire enters many subjects, yet claims none uniquely as its own.
The other "elements"—air, water, earth, even wood—have a hard materiality.
Although they also have a chemistry and are compounds of many substances,
one can pick them up, carry them to another setting, push and plunge and
pummel them. One can inflate a football with air and kick it. One can fill
a bucket with water and haul it to a field. One can dig up earth and dump
it elsewhere. One can leave them alone, untended, or seal them off and find
them again later. But one cannot pick up fire as fire. You carry its fuels, upon
which it glows or flames—you pick up embers, smoldering branches, a flaming

matchstick. Remove that fuel, and the fire dies. Shut off its air or cool it, and the fire will go out. The other ancient "elements" have intellectual disciplines and academic departments to study them. Fire does not. (The only fire department in universities is the one that sends emergency vehicles when an alarm sounds.) Fire is, in truth, not an element at all, except that its unblinking importance makes it elemental to human life.

Yet it must go somewhere. In popular consciousness, it retains a substantive identity. In most of daily life for industrial peoples, fire has vanished into, at most, ceremonial vestiges. For urbanites, who make up most of the human world, it remains real largely because it continues to burn and threaten cities and has led to institutions to prevent or control it. An institution committed to the ultimate extinction of fire, however, may not be an ideal place to nurture its identity. The goal of urban fire services is to abolish flame.

Instead, since the eighteenth century, for Western civilization the study of open burning has lodged in forestry. Others were interested—chemists, physicists, agronomists, technologists. But the discovery of oxygen destroyed fire's claim to independent identity for chemists; the laws of thermodynamics undermined it for physicists; agronomists struggled furiously to find alternatives to fire's role in fertilizing and fumigating; and technologists segregated fire's heat and light from its standing as a free-burning process. Only foresters had to confront fire as fire. A historical accident—the fusion of silviculture with state-sponsored reclamation and then an imperial outburst that reserved vast colonial estates as forests—left fire among foresters. As always, its circumstances determined its character.

How did forestry view fire? It viewed it as a disturbance. It saw something that threatened trees, degraded soils, destabilized rivers and climates. It saw a social and perhaps political failure because, for temperate Europeans, fire existed on the land only because people chose to put it there. It saw fire as unnatural and utilitarian. If there was a place for fire, that role existed because fire behaved much as foresters did—planting, pruning, harvesting. It was a technique to a shared end. All in all, however, wildfires were viewed as an index of social disorder or of nature in riot or revolt, often with human accomplices. Forestry saw a stubborn problem that begged for proper engineering.

Recent decades have reversed the particulars. Hammered by wilderness legislation, inspired by biocentric philosophies, moved by a management vision that expanded from lumber to landscapes, and penitential over past failures, American forestry, especially, has proposed complements to its earlier conclusions. Fire is now natural as well as unnatural; it is a tool that can be used to advance management goals as well as to destroy biotas; and, as fire, it binds forestry to other professions. Fire has become a common cause for wildlife, wilderness, biodiversity, range management, ecological restoration, and urban fire services. But the deeper vision of utilitarian flame remains. Fire exists as something to start or to suppress. It is a tool to use or a transcendental ecological event to leave alone. People can slash, burn, quench, or stand aside.

The time has come to recenter fire. For one thing, it is clear that lighting and fighting fire are not, by themselves, a sufficient basis for handling fire in wildlands. In fire-prone landscapes, fire control has proved self-defeating and ecologically unsustainable. But simply reversing the history of fire exclusion and ramming controlled burns into the land will not reverse a legacy of fire starvation automatically. Switching polarities may transmit electricity; it does not manage fire. Fire binges and fire busts lead only to an anorexic biota. Again, fire synthesizes its surroundings: messed-up forests yield messed-up flames. For another, the proliferating number of professions and disciplines with at least a casual interest in fire suggest that forestry, even inflated by interdisciplinary cross-links, cannot hold it all together. The prospect of an autonomous department of fire studies is improbable. The simplest solution is to relocate fire science to a more catholic discipline, biology being the most likely candidate.

The reason is simple: fire is a creation of the living world. Life produced the oxygen fire craves; life stocks and shapes the fuels that feed it; and, in the guise of humanity, life progressively oversees the kindling that sparks it into existence. The chemistry of combustion is a core chemistry of life: combustion only takes apart what photosynthesis brings together. When that reaction occurs within cells, it's called respiration. When it happens among organisms, it's called fire.

Life's organizing hand follows fire everywhere it goes. The biomass that fire consumes is not inert carbon bullion. It is (or was) alive and subject to evolutionary selection, ecological checks and balances, and considerations of scale, which range from the deconstruction of carbohydrates in mitochondria to the cycling of carbon on Earth. This hierarchy of biological organization sustains a hierarchy of fire behaviors and outcomes. Fire follows the combustibles that feed it through a trophic chain. Fire is what it eats—and only what it eats and how it eats it. Flame has no reality apart from those environs. Truly, it has the power to transmute. It can make or break those environs; but as an agent of creative destruction in nature's economy, fire can call into being or blast into oblivion only what it burns.

The human control over ignition is a stickier issue. Deep-ecology critics, especially, will quickly point out that fire predates people, that lightning and, locally, volcanoes kindle fire, that there is no reason to compromise a natural process with a human presence. Fire is natural, and people can only pervert fire's planetary purposes. But that argument misses the point precisely. One creature does possess the power to start and stop fires. If that power resided in a species of Lepidoptera or in a marsupial, and particularly if that creature set fires on an order of ten to one (or better) to lightning, no one would question the nearly total biological basis for fire. We do so only because we are that creature.

Yet more is at stake than simple ignition: fire can spread only if it has fuel. This is where humanity again flexes its firepower because we can make fuel and do so plentifully. Over most of the Earth, fire burns within the context of agriculture, of fire-fallow farming or fire-forage herding. (And if there isn't enough living biomass around, we excavate fossil biomass.) Fire burns, that is, because a creature creates the proper conditions. Other species bash and chew their environments into more suitable habitats for themselves. We shape our surroundings with fire or to make fire possible. Slash-and-burn cultivation is no less biologically driven than elephants trashing acacias or bison nibbling bunchgrass and not junipers.

By far the greatest proportion of fire on the planet and virtually all its fire regimes result from human tinkering. This fidgeting and tweaking are ecological acts. With humans, the biosphere very nearly closes the cycle of fire within itself. Not completely, because lightning still kindles fires and because, in fact, some societies have chosen to remove themselves as active agents from selected landscapes and let lightning-caused fire rule. But this self-eviction, too, results from social decisions. If we chose to compete directly with lightning, the scene would look different. We don't have to fight fire to counter lightning. It would be enough to pave the landscape into a city, for example, or plant it to genetically modified tomatoes or simply do the burning before the first dry bolts arrive. Likewise, the intellectual decision to absent ourselves as biological agents is a choice we make, not one embedded in the biosphere.

As we think about fire, so we act on it. Presently two conceptions—slogans, really—dominate discussions. One proclaims that "fire is natural." This is a truism. No one doubts that fire predates humans. No one today doubts that, in some fashion, fire belongs in nature reserves sited in historically fire-rich landscapes. A half-century ago the observation was necessary to remind federal agencies that fire exclusion could have serious costs and that not all fires were the upshot of human malfeasance and hence bad. Some came from nature and hence shared the putative goodness of the wild. The purer the nature, the better the fire. Fire-as-natural bonded fire management, in particular, to wilderness.

At first blush, this would seem a dandy argument for a biocentric philosophy of fire. The problem is the wilderness coda, for the chant quickly slides into metaphysics and an insistence that *only* natural fire belongs. This belief has implications for both fire management and ecological science. The practical difficulty is that it shuts out the prospect for routine burning—for regime maintenance by fire—as distinct from a one-off burn to "restore" a "natural" order. It ignores, moreover, the likelihood that the predisturbed state itself was sculpted by anthropogenic burning. The conceptual difficulty is that the assertion equally hamstrings ecological theory. Ecology is a historical science, much as geology is. Removing a prime cause for historical change—namely, ourselves—does not explain why the present scene exists as it does or what we should do about it. A fire ecology without humans can expound only on hypothetical or long-vanished biotas. The final outcome is not a more pristine model of ecology but a plaintive metaphysics. Of course, models are simpler if we remove people. They would be simpler still if we removed flora or fauna.

The subtext behind fire-is-natural is the quest for a Pure Wild. To prove the existence of a natural fire regime is to prove the existence of true wilderness, a transcendent nature. The arguments supporting it eventually resemble the arguments for the existence of God and, in the end, rest equally on faith. The search for a sliver of the True Wild belongs with the quest for a splinter of the True Cross. They do not tell us how we, as uniquely fire creatures, should behave. The pressing issue before fire management is not whether fire has existed previously, but in what ways—according to what regimes and by what means—and what its various presences might signify for a particular landscape today.

The second conception is that "fire is a tool." This view would seem to satisfy the need to involve human agents. If it is a tool, however, it is an odd one and a tool that exists only because of its setting. An ax exists in its own right: a fire does not. An ax cannot morph into something else as it moves from a carpenter's shop to a woods; a fire can. A grass fire can become a woods fire, and a burning swamp may metamorphose into a crown fire as circumstances allow. Clearly, people have "used" fire, and hence it may be considered a technology. But what kind of technology is it?

Several kinds. There is fire as a tool, fire as a tamed "species," and fire as a captured ecological process—or at least these are the most common conceptions. Consider each in turn. Tool fire embraces *tool* in its everyday meaning. Flame sits on a candle as a claw hammer sits on a handle. It applies concentrated heat and light. Of course, one has to feed it wax and assure it has ample oxygen, whereas a hammer does not consume the wood of its handle or demand air, but there are good reasons to treat such fires as a tool in the vernacular sense. What one wants is the heat and light. It is thus possible to substitute another tool for this one, to use heated coils and electric lights, for example, instead of flame, or to put the "flame" into a chamber that disaggregates it into the most elemental parts of combustion and then to apply its heat to power prime movers.

Tame fire works differently. It depends more acutely on its circumstances.
Such fires operate within a domesticated, usually agricultural setting: they are
the fire equivalent of domesticated species such as cows, horses, and sheepdogs.
They do a variety of tasks, much as a horse may pull a plow, draft a surrey, or
carry a rider. They may burn the pruned limbs of fruit trees, the ditches around
a farm, the fallow of a field. But their power is only as great as their surroundings, which are very much shaped by human contrivance. They are, in fact,
fire variants of kept creatures. They must be bred, selected, trained, nurtured,
housed, harnessed to particular tasks, held on a leash. They are more difficult to
substitute for. Replacing the tame fire is like substituting a tractor for a draft ox.
It can be done, but the consequences ripple through the farm.

Captured fire more resembles a caught animal such as an elephant taught
to haul logs or a grizzly bear trained to dance. Its "wild" properties are what
make it valuable. In this instance, its "wildland" or coarsely managed context
are what define the fire and sustain it. People loose it, like those cheetahs in
Mogul India trained to the hunt, and let it roam. Its success depends on timing
and of course on setting. It can go feral, quickly and unexpectedly, or turn on
its nominal guardian. Yet its value is unquestioned: it can challenge wildfire on
its own grounds, without meticulous preparation. It can substitute a partially
controlled process for an uncontrolled one. It is how, over long millennia,
aboriginal economies have turned uncultivated lands to productive purpose. But
as an ecological process it is not replaced readily. No combination of chain saws,
bulldozers, and woodchippers can do what it does. Although a technology, it
is hardly a tool in the common sense, and attempts to characterize it as such
must fail.

All of these technologies, moreover, depend not only on their environmental setting but on their relationship to humans. None of them could even exist
without a human agent. Those relationships run a gamut: tool fire is a device,
tame fire a symbiosis, captured fire an alliance—and there are others less
prominent. Fire cannot be separated as readily from its user as the naive image
of a tool suggests. It is easy to take a candle away, but less so field fires and
still less the prescribed burning of wildlands because the web of relationships
increases. Subtracting fire may be as powerful as adding it. A removed hammer
may mean a nail isn't struck. A removed field fire may unravel an ecosystem.

Instead, fire-as-biology recommends another strategy. It focuses on the overall context, social as well as biological. It envisions fire as a biotic catalyst, a synthesizer of those surroundings. It argues for thinking of fire control as a variety of biological control, much like integrated pest management. It shifts attention from mechanical acts such as starting and stopping fires and toward the interconnections that make fire possible, shape its behavior, and determine its outcomes. It forces people to accept their role as fire creatures because it is we who (nearly) close the biological cycle of burning. We are less mechanical engineers than genial hosts—warding off the unruly fire guest, welcoming the jovial, sparking a flagging conversation, dampening a smoldering dispute.

Fire-as-tool suggests that the problem is to put fire in or take it out. The solution to unwanted fire is to shut off its air supply, remove its fuel, interrupt its chain of ignition. Fire-as-natural urges, if obliquely, that people erase themselves from their heritage as fire agents. By contrast, fire-as-biology suggests that the problem is to decide what fire's context should be and then determine what kind of catalytic fire-induced jolt might best serve that setting. That fire is not merely a device to reduce fuel so much as combustibles are a means to get the kind of fire a biota requires. That our role as fire keeper is more complex than that of toolmaker because it involves ecological connections as well as tasks. That fire, for humanity, is more than a problem or a process: it is a relationship. That fire, although no longer considered an element, remains elemental.

buy
this
book,
save
the
planet

A modest proposal to decouple internal combustion from wildland fire
through the medium of books, thereby granting to fire
the fuel it needs to feed upon, the air it requires to breathe,
and the political space it demands to flourish.

In November 2000, negotiations over the Kyoto Protocol to the United Nations Framework Convention on Climate Change, which aspires to regulate the production of greenhouse gases, stalled when Europeans and Americans failed to agree on what might count as a carbon sink. Americans wanted to tally carbon stored in crops, soils, and woods, whereas Europeans refused to let Americans continue to belch out as much carbon as they could combust if only they could connive enough biotic blotters to sponge up the mess. They wanted Americans to get out of their damn cars or at least to pay the same exorbitant prices for gasoline that Europeans have to.

So the talks broke down. Americans went back to their multicar garages and revved up their SUVs. Europeans retired to their smug condescension over ill-disciplined Americans with their tendency to identify culture with extended-cab F150 pickups and all-terrain tires. One can almost see the French intelligentsia sniffing over their sagging volumes of fusty Foucault and dreary Derrida.

There may be, however, a solution. The moribund American publishing industry should turn seriously green, and not with envy. Paper is almost half carbon. Books are carbon bullion. Bookstores and libraries are carbon sequestration banks. Published books—the bigger the better—may be the missing carbon sinks the United States seeks. The more books we buy, the more miles we can drive. That's a formula that should set the publishing tachometer racing. Buy a book, save the planet.

Do the numbers. Approximately 45 percent of modern paper is carbon. A book weighing a kilo holds 450 grams of carbon. A back-of-the-envelope calculation (actually a pile-on-the-bathroom-scale reckoning) of my own humble oeuvre (fourteen books, more or less) weighs 10 kilos. This may not seem like much—I harvest a heavier bucket of citrus every morning to make juice—but everyone can stockpile books, even in Maine and Iowa, where lemons are found more often on used-car lots. If a modest one hundred thousand people bought my entire list, America proudly would have amassed a million kilos of processed carbon, roughly what the residents of Phoenix release every morning from gasoline just to get to work. That trees are minced to make paper is irrelevant—or, rather, the whole point. The trees grow back. The books stay.

It's a virtuous circle. The more books the public buys, the more robust the publishing industry. The more vigorous publishing becomes, the more people will be enticed to write and the more they will make from it. A proper government would promulgate policies to encourage such a process. A proper government would enact carbon tax credits for those who buy the goods and dole out carbon subsidies to those who produce them. Houses would then swell with shelving and groan under the weight of collected works. Book superstores would sprout like mushrooms. Libraries would proliferate like Circle Ks. The publishing industry's chronic lament that it is economically marginal would vanish overnight. The fuel industry would buy up publishing companies to complement its existing distribution networks. Gas stations would include self-serve, buy-at-the-pump paperback dispensaries. Like a ringed benzene molecule, an industry would close the cycle of carbon spewed out and carbon sopped up.

The real beauty of the scheme, however, is that no one actually has to read the books. In fact, there is a case for leaving them in their shrink-wrapped cellophane (which adds its own, admittedly infinitesimal, carbon increment). Here is an opportunity for deep(-time) discounting. Sure, the Europeans will squeal, but they hardly can object to Americans' amassing of both carbon and culture. After all, the French have managed to make their own "cultural" creations exempt from World Trade Organization rules. Even they can't bare-facedly denounce a massive expansion of U.S. book publishing as harmful to either trade or ecology. If enough people bought it, for example, Al Gore's *Earth in the Balance* actually might do some good, without the bother of having to follow him from one belabored point to another. In France's case, more books bought and fewer seriously read might help the world get on with its business.

The open road and the unopened book—it's America. We can fill up our tanks with all the gasoline we want so long as we fill up the trunks with books. Let the Kyoto negotiations begin anew. Let the French suck eggs.

bearings,
corrected
and
otherwise

hominid
hearth

Africa is a fire continent. Even discounting the Sahara—a desert nearly as large as Europe—Africa sparkles with more routine fire than any other landmass. The significance of African fire, however, derives not merely from its extent but its antiquity. Anthropogenic fire originated here and has resided here longer than anywhere else. These two facts, Africa's antiquity and its size, make African fire equally unique and undeniable. Its fires calibrate wonderfully Africa's natural history.

Begin with geography. Fire flourishes because most of sub-Saharan Africa has an environment to sustain it. There exists, first, a chronic rhythm of wetting and drying. Seasonality follows a cadence of rainfall, not temperature. Wet seasons grow fuels, dry ones ready them for burning. Across this annual rhythm, longer waves of drought and deluge rise and fall. The formula is ideal for fire, and the onset of the rains typically brings scattered thunderstorms laden with dry lightning. Far before humans arrived, fire thrived, and it will continue to do so—bar only the driest and wettest sites—if people leave.

Other features of the African scene encourage burning. Africa is the highest of the vegetated continents, which helps explain its extraordinary mineral wealth (its exposed kimberlite pipes and enriched copper deposits, for example), but that same process of deep erosion and leaching has left much of the continent with impauperate soils. Some volcanism along the Rift Valley has renewed nutrients locally, and river floodplains receive regular loads of fresh alluvium, but most of the soil, by the standards of the Northern Hemisphere, is relatively indigent. Anything that can move precious nutrients through the biota will claim an important role in nature's economy. That, fire does.

Equally, Africa's fauna matters enormously. Nearly 75 percent of its megafauna survived the great Pleistocene kill-off, far more than on any other continent. These animals profoundly shape the flora, which is to say, the fuels on which fire must feed. In fact, the two combustions compete, the slow combustion of respiration with the fast combustion of fire, metabolizing animals with the flames of burning biomass. The interactions are difficult to track precisely, for animals, plants, and fire make a kind of ecological three-body problem that can be understood only approximately and locally. In general terms, the relationship depends on whether the land is prone to fire or not. If not, the animals are often vital in agitating the fuels into forms that can burn—bashing over trees, for example, opening canopies, stirring the surface vegetation. If the beasts vanish, the conditions for burning may vanish with them. If, however, the land is fire prone, then removing megafauna can liberate fuels. Grasses and scrub that otherwise would be eaten can be burned. The survival of Africa's megafauna has meant, overall, an increase in the land area subject to fire. Probably the striking edges that abruptly separate burning veld from unburnable forests are a border erected by fauna as well as by flame.

The biota of sub-Saharan Africa, then, has long adapted to the burdens of drought, grazing, browsing, impauperate soils, and, in most sites, fire. Its flora tends to exhibit a suite of traits that help accommodate this suite of stresses. Such a biota can respond readily to more fire or, more properly, to a shift in fire's regime, for the issue is not whether fire, stark and singular, comes or not, but what patterns its arrival and departure assume. Its *regime* describes fire's typical size, seasonal timing, frequency, intensity—the patches and pulses by which it manifests itself. (A regime, remember, is a statistical concept. Individual fires are to regimes as storms are to climate.) In fire-prone lands, these characteristics result mostly from the configuration of ignition. And that is why hominids matter and why Africa justly can claim special standing for the insights it provides regarding the meaning of the human species monopoly over fire.

The capture of fire by early humans marks a profound moment in the natural history of the Earth. So fundamental is combustion to life on the planet that the ability to apply flame more or less at will conveys an unprecedented power on any species that holds a firestick. The history of humanity as an ecological agent can be tracked by our ability to expand the dominion of fire and our willingness to apply fire to whatever purposes we conceive.

Probably the earliest hominids could carry and nurture fire but not kindle it on command. *Homo erectus,* for instance, likely could tend fires seized from natural sources or acquired from other bands of hominids. But the instruments for starting fires such as striking flints, abrading wood, and twirling drills derive from a toolkit not developed until later times. With *Homo sapiens,* however, the capacity to combust becomes as common as the ability to drill holes in bone or knap obsidian into spear points. From this instant on, the dominant source of earthly ignition has been anthropogenic.

The ability to create spark, nonetheless, is not synonymous with the ability to propagate fire. The human firestick is only as powerful as its setting. If the land is not ready to burn, no shower of sparks, however torrential, will set it aflame. A match won't set an empty fireplace ablaze. What matters is the receptivity of the land. If it is fire prone, then people easily can assume control over its regimes. The reality is that anthropogenic fire competes against natural fire, usually by burning early in the dry season or otherwise changing the timing of ignition. Its power amplifies where the megafaunal population has thinned and thereby has liberated more fuels to feed the flame.

Still, even in a generally fire-favored continent such as Africa, this leaves vast patches of land untouched by flame of any kind. For humans to seize the commanding heights of nature's economy, they need to control fuels as they do ignition. This, from the perspective of fire history, is the meaning of agriculture. Farmers and herders could apply axes and mauls, tooth and hoof, to crunch, fell, grow, dry out, and restack biotas so that they might burn when they would not do so unaltered. More fuel means more fire.

As agriculture began its slow six-thousand-year colonization of Africa, it refashioned the geography of fire in two ways. First, it remade many aboriginal fire regimes. Fire's pulses beat to new rhythms, fire's patches propagate to new purposes. Agriculture disciplined flame into the social order of farming and herding, while crowding hunters and foragers to more marginal sites. The fire options were not simply between nature and humanity, but between nature and several variants of anthropogenic burning. And second, agriculture pushed fire into places such as shade forests that it previously had been unable to penetrate. The expansion of Bantu speakers is the best-known and most perva-sive example, and because that colonization was accompanied by iron hoes and axes, it hauled across these landscapes the forge fires of metallurgy (with their inexhaustible craving for woody fuel). The overall geography of African fire expanded enormously, and it altered its character. African fire became increasingly obedient to the whim and will of human culture.

The character of cultivation derived from the catalytic power of flame. Few natural sites encouraged sedentary agriculture—and those tended to cluster along rivers or deltas, where water could take the place of fire as a disturber, as a means to purge and promote. Wet epochs could make burning difficult; drought could starve fire of its fuels. The solution was to move through the landscape, to rotate the farm through the countryside, to allow fuels to regrow, and then to reburn. In this way, people could exploit the basic principles of fire ecology, which underwrote this system of fire-fallow farming. For a year, the site was fumigated and fertilized and open to the cultivation of exotic plants; within two years, the indigenous biota would reclaim the site, choking it with weeds and (from an agricultural perspective) with vermin. The abandoned plot would resprout, until, at some time in the future, it would be slashed, dried, and burned again. The celebrated *chitemene* system common in southern Africa is a textbook example.

Yet limits remained. Not every biota could be minced into combustibles and burned when convenient for agriculturalists. Megafauna competed with domesticated livestock. So did microfauna in the form of diseases that plagued people and their servant species, notably the tsetse fly and trypanosomes that long stalled settlement. And although fuels swelled from the hoofs and axes of agriculture, they remained bounded by the larger properties of soil and climate. One could not convert more biomass into fuel than nature, even assisted, could grow. The realm of fuels spread out but never transcended the green shackles of its ecological roots.

One also could add to this broth of historical factors the commercial isolation of Africa, which buffered against stimulants from the outside world. Sahara sands shrank commerce northward to a trickle. Rapids and falls shut down river traffic. Good harbors were rare, and long-range maritime shipping sparse, save along the eastern flank, where Swahili settlements sprang up to integrate parts of Africa into the commerce of the Indian Ocean and the *ecumene* of commercial Islam. But until far-voyaging Europeans ventured along the Atlantic coast, there was little prod elsewhere. The geography of African fire responded to the internal dynamics of migrating peoples, in particular swiddeners from West Africa and pastoralists pushing through Ethiopia, both pressing southward. Even after European outposts dotted the littoral, they remained trading factories. For four hundred years, Europeans clung to the coast like barnacles. They traded; they didn't—couldn't—move inland; diseases alone stalled any serious demographic surge. Their impact was indirect.

When they finally did push inward, they did not colonize by massive immigration, save spottily in far southern Africa. The principles that governed African fire ecology endured. Francis Bacon's dictum held as true for fire as for other technologies: nature to be commanded had to be obeyed. Fire persisted, for both fire-wielding peoples and fire-adapting biotas both commanded and obeyed. Kingdoms rose and fell with the same kind of patch dynamics that ruled fire ecology. What changed were the particular expressions of those principles as populations became scrambled and trade routes fell under new jurisdictions. Even the horrendous collapse of native societies in the late nineteenth century—the crash of populations (of people, livestock, wild ungulates) precipitated by the spread of rinderpest and market hunting; the crushing long drought teleconnected to millennial El Niño–Southern Oscillation events; the pathological disruption wrought by the final spasms of the slave trade, war, and colonial conquest—yet remained within the larger span of African experience. The ecological aftershocks, although unique in details and horrific in extent, were broadly predictable. They would move the circle of ecological transactions in and out, like a biotic bellows. But the basics endured, and the ecological circle of fire remained unbroken.

The circle snapped only when European imperialism became the vector for industrialization. From the vantage point of fire history, industrialization refers to the wholesale combustion of fossil biomass. The Industrial Revolution fundamentally reconfigured human firepower by expanding stupendously the stocks of combustibles. For practical purposes, they became unbounded, and the ability of anthropogenic burning to affect the landscape, both directly and indirectly, promised to shatter the confining rings and rhythms of ecology. Industrial fire not only competes against other combustion but—at least momentarily—transcends it.

The general properties of industrial fire are poorly understood, but these features seem universally true: Where fire exists as a "tool," industrial combustion can substitute readily and does. Where fire exists within an agricultural context, a resort to fossil fallow eventually shoves aside the old practices of creating fuels out of living fallow and then burning them. In fully industrial nations, free-burning fire survives largely in nature reserves or, temporarily, on abandoned agricultural fields. The Earth is dividing into two great combustion realms: one dominated by the burning of living biomass, the other by the burning of fossil biomass. Few lands hold both types of burning or do so metastably.

With scant exceptions, sub-Saharan Africa belongs with the first group. It persists as a place of open flame sustained by aboriginal fire practices or by fire-fallow agriculture and pastoralism. A few enclaves have escaped, of which portions of South Africa, rich in coal, are primary. How Africa might industrialize is unclear. Around the Earth, there seems to exist a pyric equivalent to the demographic transition: a troubled period when old fire practices live on and new ones flourish, causing a massive swell in the overall population of fires. Uncontained fires breed like flies. Once-docile fires turn feral. Fires seem to overwhelm the land. Eventually this phase passes. Industrial fire practices substitute for their predecessors; the fire load declines; combustion throbs within machines, outside direct contact with the land. Society absorbs and contains combustion as ancient organisms did respiration. But the era of exploding fires often brands itself into memory, especially for elites. That industrialization coincided historically with an era of colonial conservation no doubt sharpened the critique of open burning.

All this is happening to Africa, though slowly and spottily. Countries such as Nigeria and Angola hold immense reserves of petroleum, yet they export those fuels and continue to rely on open flame for daily life. That disconnect is characteristic. Rural fire is everywhere difficult to extinguish. Even in places such as Europe, which lack a natural basis for fire, it loosened its grip reluctantly. It took more than sixty years in the United States to unshackle the most fire-flushed rural landscapes. One complicating problem in Africa is that agriculture cannot move easily into a fossil-fallow regime. Besides, rural fire is not ultimately a fire problem but a condition of rural life. To eliminate it, Africa will have to become something very different from what it is now. A reform in fire practices can only accompany a general reformation of society.

Even so, fire will flourish. Favorable geographic conditions will endure; industrial peoples, typically urban, tend to reserve swathes of land as nature reserves, and on these reserves, in Africa, fire will persist. The abolition of fire is, in any event, an unnecessary, problematic, and ironic ambition. Fire will endure in Africa because Africans will wish it to. The real issue is what kind of fire they want, to what ends they desire it, and by what means they might allow such fires to do what fire ceremonies universally proclaim fire to do: to promote the good and purge the bad. The fire in the hominid hearth is unlikely to expire soon.

fire's lucky country

The bushfires have returned. Declarations of emergency, water-dumping helicopters, mass evacuations, mobilized fire brigades, flames engulfing houses and washing over national parks—in a broad arc throughout the southeast, bushfires rang out the old calendar year and rang in a searing 2002. But this is not news. Almost every year, somewhere, Australia's bush burns, every decade or so spectacularly. This has gone on spasmodically through geologic time. For probably more than the past forty thousand years fire has been almost constantly on the land, so much so that it has become as indomitable a shaper of Australian ecology as soils, climate, and the raw genetic stock of the biota. One might as well try to explain the Australian scene without deluges, droughts, cyclones, leached soils, and marsupials as to explain it without fire. An outburst of flame is not news.

The news is the remarkable removal of fire over the past century, not merely from the land but from the hand. There is less flame overall, fewer controlled fires set by people, and more damaging conflagrations. The real news occurred six months before Black Christmas, when a public seminar on fire science sponsored by the Western Australia Conservation and Land Management Department (CALM) was forced into postponement, apparently until it could be properly vetted politically. Between them, the wildfires in the southeast and the problematic fire symposium of the southwest, neatly bookend the contemporary state of Australian fire. The real story is Australians' continuing struggle to reconcile with their burning bush.

Fire has its reason and its season. What binds the two is the rhythm of wetting and drying. Very little of Australia lacks such a biota-fracturing rhythm, which is why very little of Australia fails to burn. But wetting and drying, by themselves, do not cause land to combust spontaneously. That demands some spark of ignition, a kindling. In nature, lightning does this job, and lightning's lottery is one reason why natural fire is lumpy. Lots of potential fuel remains unburned; lots of sparks expire without spreading. What nature's fire economy needed was a broker who could match flame with fuel. That task fell to *Homo sapiens*. With people, spark is almost constantly on the land. Since the first colonization of *Homo*, most of Australia's lands have been burned routinely, in ways both intricate and coarse, and its fire regimes shaped by the restless hand and contriving mind of Earth's keeper of the flame.

Still, there were limits. Not every place underwent a wet-dry fracturing, and not every place that could conceivably burn did. Not all burning was deliberate: fire littering proliferated alongside fire foraging. Because fire followed fuel, burning was as patchy as one year's flush growth and another's withered scrub: little burning occurred with metronomic regularity. But, in general, what could burn was burned, and the biota accommodated itself to the particular rhythms fire manifested on the land. This has gone on for tens of millennia. Whether or not fire had to be, it simply was. It was there, like rain and sun and dry winds. Flora and fauna so adjusted to its presence that fire's removal could be as disruptive as its appearance.

The chief bound on Aboriginal fire was that one could burn only what nature itself made available. The ability to make fuel as readily as spark is a fire characteristic of agricultural societies, and it is why Australia's rivals as fire continents, especially Africa, sparkle with the flames of farmers and herders. Australia, however, remained an Aboriginal fire continent, as distinctive in its fire regimes as in its marsupials and mammalian pollinators. Why? If one considers agriculture as a fire-catalyzed ecosystem, the explanation is simple. It wasn't needed. Sufficient fire already existed: the conditions for burning were abundant without the added labors of slashing and fallowing. As Rhys Jones famously phrased it, "firestick farming" was already pervasive. There was little compelling reason to expand fuels because most of the landscape was already accessible to fire. To demand further accessibility depended on the requirements of exotic cultigens, which did not arrive until 1788. Surely the absence of agriculture in Aboriginal Australia, which has puzzled anthropologists and agronomists for years, requires an explanation more complex than this, yet it is doubtful that any explanation can succeed without an appeal to fire ecology and may very well need to begin with the fact that the land already simmered with controlled flame.

All this of course changed with European colonization. For fire's ecology, what most distinguished British settlers was their capacity to wrench the landscape into more combustible forms. They could alter fuels on a continental scale, both directly by cutting and planting and indirectly by importing fire-thirsty weeds and encouraging their livestock to crunch through landscapes. The fine fuels that had fed the fast combustion of flame now often fed the slow combustion of metabolizing sheep and cattle. The new settlers moved the fulcrum of fuels, and by so doing elevated the firestick's leverage.

Like a colossal kaleidoscope, the landscape mosaic was given a rough twist, and a new order, or rather a succession of orders, began to tumble into view. Fire persisted, but was either loosed into the ecological sprawl of colonization or leashed to the fields and paddocks of an emigrant agriculture. Bushfires flared in the wake of prospectors, loggers, and selectors, or, once the land was cleared, they settled into the semidomestication of stubble-burnt fields. The bushman, notoriously, emulated the firestick habits of the Aborigine, kindling billy fires wherever he stopped and inscribing new patches and pulses of burning.

Yet almost as quickly as Aboriginal fire began morphing into agricultural fire, a further order of burning appeared. Industrial combustion—the burning of fossil rather than living biomass—overcame the limitations of agricultural fire, which knew strict combustion boundaries because it could not burn, repeatedly, more than nature could grow or be coaxed to grow. By combusting fossil biomass, however, industrial fire tapped the immense landscapes of geologic time, whole new continents of potential burning. This wave of fire colonization has barely reached adolescence. Its fire ecology is as obscure as it is profound.

It is known, though, that this new fire immediately began competing with the others: the Earth is rapidly fissuring into two great combustion realms, one powered by fossil biomass, the other by living. Confronted with the Big Burn of industrialization, much of the old order of fire has disappeared. A good bit of burning—domestic, technological—was cast aside in a rush for more precise substitutes. This process was simple and benign. Electric lights could nicely replace candles and wood tapers; automobiles, oat-devouring horses; gas furnaces, wood-burning hearths. But the process did not cease at the domestic threshold or at the borders of the city-states that increasingly housed the human population. Along the way, industrial fire practices have sought actively to suppress open flame nearly everywhere, and the attempt to do so has rippled through the biosphere, upsetting biotas as surely as fossil-fuel greenhouse gases have perturbed the atmosphere.

Australia is a firepower. Although Earth is a fire planet, it has its hot spots, and Australia is one. Why? Because, first, Australia has, as a tenet of its geography, the natural conditions for fire and, from its human history, a long legacy of burning. Because, second, the nature of European colonization left large patches of Australia emptied, for a time, of its indigenous peoples and not yet reclaimed by new immigrants. It possessed large extents of relatively wild bush that would remain under the jurisdiction of the state as Crown lands or, more particularly, as land reserved for various public purposes such as forests or parks.

They would not pass into private hands. They would remain as public entities overseen in the name of the common good. And they required some agency of the state to administer them. The practical reason was that local communities had, by and large, proved incompetent to shelter landscapes on behalf of extralocal values and especially to shield them from larger market forces and political pressures. Villages and nomadic tribes had scant incentive to restrict burning or land clearing, for example, in the interest of stabilizing climates, and too often they collapsed in the face of economic goads to scalp forests or to replace woods with plantations. Some larger institution had to intervene between the land and its exploitation. An emerging philosophy of conservation argued that only national or imperial institutions could satisfy the burden of those ends, while an obsession with the putative relationship between woods and waters (especially between deforestation and drought) suggested that forestry might be the appropriate means. The outcome was a global era of state-sponsored forestry.

The practice of reserving wooded lands was as common a feature of

European imperialism as military cantonments and missionaries. The scheme worked, reasonably well, only where the lands remained genuinely uninhabited. In nations such as India, Ghana, and Cyprus, where the natives persisted, the lands were never truly vacant, and the reserves could never isolate themselves from rural life and rural fire. But in the United States, Canada, Russia, and, of course, Australia, they could, within limits. Such nations were able to resist the implacable pressure for industrial extinguishment. At least in patches, some quite vast, they held fire as open flame; they sustained fire in its free-burning form.

The possession of such fire-prone public lands places Australia into a small club of cognate firepowers. There are lands that have as much or more fire than Australia, but they thrive within an agricultural matrix—Africa and vast chunks of South America, for example. Formally reserving the land, however, spared it from agricultural reclamation, urban sprawl, and combustion conversion by a remorseless industrialization. It would remain as bush. And this being fire-flushed Australia, the bush would provide a permanent habitat for fire. An enduring bush ensured enduring bushfires.

This has two consequences. One is a dilemma that all the members of the public-lands club have confronted. Because such lands, if fire prone, will burn, either you convert them to something less combustible or you do the burning yourself or you accept that nature's fire economy will careen through fire boom and fire bust, often violently. Moreover, in most instances the land is not adapted to natural fire regimes but to those laid down over thousands of years by people. Abolishing fire is not an option: the old internal baffles overgrow, the fuel stacks up, patches intertwine, and the fires become more savage. Apart from control issues, megaintensity fires can degrade ecological values and scour out biodiversity. A landscape used to small surface fires, for example, will not take kindly to ravenous, forest-stripping conflagrations, any more than a landscape accustomed to rainfall distributed evenly month by month will accept readily a shift to gully-washing deluges that occur only once or twice a decade.

The second upshot is that the state becomes committed to fire management, including fire research, which helps account for the oddly disjunctive geography of fire and fire science. Most of the world's fire resides in nonindustrialized lands such as sub-Sahara Africa; most of the world's science belongs in Europe and European-colonized, industrialized countries. The places where the two overlap are, not surprisingly, that small commonwealth of public-land countries, among which Australia is prominent. Thus, no student of bushfires can avoid Australia. Its fires are everywhere: live flames on the ground, printed flames in the literature, an indelible part of a global discourse about how fire behaves and how it interacts with ecosystems and societies. Australians have contributed conspicuously to all areas of fire science and have excelled at field studies, in effect adapting rural traditions of burning off to the intellectual discipline of science.

Australia thus controls a hefty portion of the global economy of fire. It has plentiful bush, suitable conditions for burning, active fire science programs, and abundant examples of natural, anthropogenic, and industrial combustion, all quarreling among themselves, fire's intractable three-body problem. But it also has a distinctive fire character of its own, and it is worth isolating some of those features.

- Almost uniquely Australia holds a record of Aboriginal fire practices. The abrupt encounter between Enlightenment Europe and Aboriginal Australia furnishes the richest testimony anywhere about how fire functions in Aboriginal economies. Unlike missionaries in the Americas, for example, Enlightenment explorers were keen observers of natural history, and they often met Aboriginal fire before trade, disease, livestock, and varieties of agriculture altered traditional regimes. This fact holds more than academic interest because the Aborigines' undeniable fire presence offers a Third Way in fire strategy between the European obsession with the domesticated landscape, for which fire exists as a tool, and the American obsession with wilderness, which prefers only natural fires. In Australia, the argument for anthropogenic burning cannot be evaded. The fire regimes encountered by Europeans were shaped, if in various ways, by human agency, and the removal of that agency may unravel landscapes. No other Third Voice of fire strategy exists. If it chooses, Australia can stand alone—not as a European colony, not as an American clone, not as an Asian echo, simply as Australia. Its bush is neither garden nor wilderness.

- Its reliance on a citizen-militia of fire brigades makes Australia the Switzerland of firefighting nations. The model is far better suited for developing countries than an American-type standing army for fire control, and as governments continue to devolve, the Australian exemplar likely will spread even to nations that traditionally have depended on a command-and-control fire-suppression force.

- In keeping with its political temper, Australia's fire institutions are remarkably diffuse. Even with respect to research, there exists no single national focus. Although the Commonwealth Scientific and Industrial Research Organization (CSIRO) has sponsored, spasmodically, large-scale research programs, most of what is known about fire has come from states, universities, and individuals, a corroboree of interested parties quite unlike the scene anywhere else. The amount of research is astonishing—one might rightly consider it in aggregate as one of Australia's major contributions to international science. But it has not been packaged into a national program or even orchestrated into a recognizable national style. It boasts no single masterpiece, nothing equivalent to, say, the Sydney opera house. For that reason, however, it appears less vulnerable to political fads. That the Australian fire community has become a major institutional sponsor for the *International Journal of Wildland Fire* is a good example of how it projects Australia's presence beyond its shores.

- Far more than any other industrial nation, Australia has held to a tradition of controlled burning. One explanation is that Australians had no alternative, in part because no state institution was sufficiently powerful to impose or to attempt to impose its will on the land and in part because bushfires are ineradicable in an unyielding land given to droughts and dry winds. But much as rural Australians adapted the firestick habit of the Aborigines, so later foresters accepted—if reluctantly in many instances—the value of proper burning.

Here is where Western Australia, in particular, has contributed to this global discourse. Public foresters found ways to integrate controlled fire into their cycle of land use, including logging. After the 1961 season, their experience became one of the cornerstones of a distinctively Australian strategy of bushfire management that sought to discipline rural burning with physical science and to muster institutions for the control of fire. In the mid-1960s, aerial ignition in Western Australia made this antipodal revolution practical and granted to New Australia nationalists a bold countersymbol to North American slurry bombers, those Yank tanks of international fire suppression.

This legacy has aroused a number of Australian critics, notably environmentalists. They often see fire as an enabling device for other practices they dislike, such as logging. They may see any human intervention as unwarranted and as a challenge to deep-ecology aspirations for nature reserves. Many exhibit a curiously binocular vision—in part derived from British sources, for which fire is only an expedient implement and an unnecessary evil; in part transplanted from American inspirations for a transcendental nature, in which fire may be welcomed, but only if it is wholly natural. Industrial societies much prefer covert combustion to open flame.

All this strikes an outside observer as curious. Although CALM's exuberant program of prescribed burning may alarm, say, the Western Australia Forest Alliance (WAFA), most of the world's fire community looks covetously at Western Australia's success in keeping some reasonable rhythm of fire on the ground. Most American fire officers would give their eyeteeth for circumstances like CALM's because they have learned to their grief that burning is not an easily reversible process. One can't suppress and restore fire as one can flip on and off a light switch. Reinstating fire in fire-famished places such as the American West is akin to reinstating a lost species. It demands considerable work to create a suitable habitat; it is costly, laborious, controversial, and dangerous. Prudence would argue not to remove fire until the political planets and social stars come into alignment, and you can decide how to restore flame, but to keep some manageable fire on the land until you figure out how, in select ways and places, to remove or amplify it. Its fire legacy means CALM and other parts of Australia that have preserved controlled burning can adjust their field programs more easily to the changing values of Australian society.

What, then, is the future of Australian fire? It is what Australians choose for it. Contemporary Australia does not have a fire problem: it has many fire problems. Proposed solutions will be various, and they will have meaning only in terms of specific sites. (The only good advice is that which fire management shares with real-estate speculation: location, location, location.) Some places require more fire, some less, some a change in regime. Overall, Australia's fire difficulties are those common to industrial societies and to nations with extensively reserved bush in historically fire-rich landscapes. Like those others, Australia's fire future will involve the braiding of three grand narratives: an industrial narrative, an imperial narrative, and a national narrative.

- *The industrial narrative.* The Earth will continue to segregate into competing combustion realms. Yet no one understands, in anything like its full ramifications, the impact of industrial combustion. The competition between industrial fire and its biomass-burning rivals goes far beyond issues of air quality and greenhouse gases, for industrialization cascades through and refashions whole landscapes. Open burning only disappears where industrial fire proposes substitutes or seeks out its extinction. This is the deep driver of Earth's contemporary fire economy. Fire has the capacity to make or break environments. Industrial fire is doing both, and we hardly understand how.
- *The imperial narrative.* The habitat of free-burning fire—of fire science and fire management institutions and of most fire controversies—is the domain of public lands. This dominion was an accident of European expansion; all public lands date from less than 150 years ago, and most from less than 100 years ago. Yet the past 50 years have witnessed a global process of *decolonization.* This has extended to the reserved lands and to the institutions created to administer them. In particular, we are witnessing the end of the life cycle of state-sponsored forestry. Devolution, the reclassification of public lands, recision to indigenous peoples, privatization—all are profoundly affecting the potential habitats of fire, the means by which they are managed, and the ways by which fire is understood.

The fine-tiled fire mosaics that characterized many preindustrial landscapes existed by virtue of labor-intensive effort. State bureaucracies are unlikely to replicate such scenes because they can never hire enough personnel or oversee their practices in sufficient detail. Instead they must fragment their large holdings into smaller shards, and they must appeal to local communities to assist. Some such communities will live on the margins of otherwise vacant public lands; others may have to be created. One can imagine, for example, "communities" in the form of nature-interested nongovernmental organizations (NGOs) that might supply labor seasonally or to special sites, re-creating a kind of contemporary nomadism that would allow human intervention of particular sorts to particular places. The present proliferation of NGOs is creating the political equivalent of a black market, obviously filling needs that formal institutions have failed to satisfy, yet doing so in ways that are socially unaccountable. Regardless, the global era of state-sponsored forestry as the grand overseer of such lands is fast passing. It is unclear how fire will survive the continued transformation; the likeliest outcome will be a further reduction in open burning and a shrinkage of serious fire science.

• *The national narrative.* Any fire practice—any policy, any research program, every fire regime—is a cultural artifact. It reflects a negotiation between what a society wants and what its land will accept. The context that gives fire its character, the surroundings that it synthesizes, are cultural as well as natural. One can analyze fire regimes as one might scrutinize a society's architecture, literature, or political institutions for what they reveal about that culture because ultimately the discussion concerns a moral universe of values, beliefs, judgments, and identities, about who we are and how we should behave. This is true for all peoples, but it is a more powerful truth for Australia because there the debate about fire has fewer baffles and buffers to dampen it from the deep reefs and storms of cultural meaning. Very quickly, discussions about what fire to apply and withhold, by what means, and to what ends founder on shoals of identity politics.

Australia has always been urban: from its British founding, it has centered in city-states, colonial capitals like miniature Londons. Yet it had sought a national identity in the bush; celebrated the bushman; made the encounter between Australia's unique biota and its émigré Europeans the theme of a national epic. That impulse has remained. What has changed is that the action and attention no longer center in rural hinterlands but in the more putatively pristine bush untouched by rural economies of grazing, farming, and mining. The Dead Heart of Australia has gone from being a symbol of the emptiness of Australian life to a throbbing emblem of its mythical essence. The Red Centre has greened. The rest of sun-baked Australia has greened with it.

This has consequences for fire. The urbanites of New-New Australia have little personal contact with flame. They no longer cook over it; clean up paddocks, orchards, and fields with it; clear new lands with it; or protect themselves by preemptively burning off. The rural bush is becoming an urban bush, subject to a fire protectorate that projects urban and industrial values. The sprawl of cities has created a landscape hostile to burning: the only fires experienced tend to be those that ravage this exposed and overgrown fringe. Equally, traditions of burning off seem to exurbanites of a piece with the market hunting, pest poisoning, logging, plowing, and paving that had inspired a Great Extermination of indigenous Australia. New-New Australians tend to want the wild bush and the urban, but little in between. That threatens to wipe out the grand domain of anthropogenic fire. A place that has burned for eons, that has long claimed center stage in the drama of humanity as a uniquely fire creature, proposes to extinguish the flames as unnecessary, harmful, baleful, embarrassing.

Yet there is—potentially—a way through the smoke. It comes by recentering fire as a biological phenomenon in which fire assumes all the nuances, diversity, and dynamism of the living world that sustains it. Such a view would suggest, for example, that instead of using fire as a means to reduce fuel, we manipulate fuel to get the kinds of fires we want. It suggests that fire is less an exogenous disturbance—an alien intruder—than a fundamental feature of living landscapes, an interactive reaction. It characterizes fire as an ecological catalyst, an agent of creative destruction.

It argues, too, for more rather than less human agency because we are the means by which the biosphere wrested control of ignition away from lightning, and we very nearly have closed the cycle of fire's biology. Other species bash over trees, dig holes, eat plants, hunt animals, but we alone burn. We alone express our ecological agency through fire. We shouldn't hand the task back to nature because nature gave it uniquely to us. We remain the keepers of the flame, however slovenly we practice our craft.

This is a master narrative in which Australia very much matters to the rest of the world. With respect to fire, the tyranny of Australia's cultural and geographic distances becomes vanishingly small. Third-Way Australians still have the chance to take fire tracks that have closed for the rest of us. And that is why Australia remains fire's lucky country.

old
fire,
new
fire

In 1998, the fires were everywhere. They started two months earlier than normal and continued two months later; they burned in every state; they flooded the sky with a pall of smoke that was swept in a colossal gyre into Texas, Louisiana, and Florida. They burned pine plantations and pastures in the high basins of Puebla. They incinerated great holes in the green woods of Lagunas de Zempoala National Park in Morelos. They fired *barrancas* and mesas in the Sierra Madre Occidental. They burned chaparral in Baja California, tropical woodlands in Yucatán, savannas in Veracruz, scrubby *matorral* in Nuevo León, parched *pastizales* in Chihuahua, oak and yucca hillsides in Tlaxcala. They branded virtually every *cerro* in the Valley of Mexico and burned over many. They gorged on the putatively immune cloud forests of Oaxaca and Chiapas. They so polluted Mexico City that authorities grounded 40 percent of automobiles. They tied down 139,000 troops of the Mexican army and scores of aircraft. They killed crews sent out to control them—19 campesinos in a canyon near Ixmatitlán, 10 soldiers in a helicopter in Tamaulipas, firefighters, volunteers, more than 60 in all. They burned in record numbers (14,136 fires), over record areas (540,859 hectares), at record costs ($33 million, plus U.S. aid), with record violence. They burned themselves into the political consciousness of contemporary Mexico.

Fire is nothing new to Mexico, of course. Its presence is as old as Mexico's volcanoes, the central *meseta,* summer storms, and stress-toughened biota. The environmental requirements for fire are simple. There has to be enough moisture to grow fuels, a long enough dry period to ready them to burn, and a source of ignition; lightning works nicely. It helps if the landscape is broken, full of peaks and canyons, a terrain that can segregate a thunderstorm's rain from its bolts, that helps ensure some part of the landscape is ready to burn. It helps, too, if a hierarchy of wet-dry rhythms and disturbances is present, if there are episodic droughts, wind storms, and eruptions to break down, rearrange, dry out, and otherwise render surface vegetation into fuel. All this Mexico has and has had since before the climatic revolutions of the Pleistocene.

Where fire is plentiful, biotas exist that have adapted to it. What allows a plant to survive fire typically also serves against stresses such as drought and browsing; fire, after all, is an interactive event, rarely a presence by itself. The norm is that a suite of traits adapts to a suite of stresses. But many features seem remarkably specific to fire. There are pines that hold serotinous cones, which liberate their seeds when heated; that can resprout from the root collar; that can send out new branches even after all needles have been killed by scorch or even burned away. There are oaks that regenerate from stumps, branches, and trunks, whose new growth can carpet seared hillsides like dandelions. There are bunchgrasses, burned annually, that glow with lush greenery as they reclaim blackened slopes. There are ferns, flowers, shrubs, cacti, yuccas—all of which shrug off a passing flamefront as they would a rain shower. Mexico's fires have survived extinctions and recolonizations. They have been on the land for a very long time.

Mexico also has had humans for at least ten thousand years. They have exploited its potential for fire and their own species monopoly over flame to shape a livable landscape. Unlike other ignition sources, anthropogenic fire became a more or less permanent feature. It burned day and night, winter and summer, wherever humans traveled, rested, or sought to reside. Fire was integral—an indispensable catalyst—to whatever they sought to make of the landscape. The torch served as a crowbar to lever out unusable tiles from the Mexican mosaic; the furnace became a crucible to smelt down and reforge rocks and forests; the controlled burn morphed into a vast flaming broom to sweep away the biotic dust and cobwebs from this ecological household. But a ceaseless fire demands continuous fuel as well as a quenchless spark. Where the climatic rhythms made fire naturally rare—places that remained stubbornly and singularly wet, dry, stable—humans had to turn to another technology to create the fuels their fires demanded. They found it in agriculture.

Farming is a controlled disturbance that replaces, for a while, one biotic assemblage with another. It is easy to dismiss agriculture's fires as irrelevant or wanton, the ecological equivalent of burning garbage. In fact, fire is fundamental. It is pointless to slash unless you can burn and useless to plant unless the soil is rich with nutrients, open to sunlight, and purged of pests and weeds. That, of course, is what fire does. Felling and burning is a way of forcing nature to yield conditions that it normally denies. With fire, it is possible to transplant cultigens such as maize, beans, and squash for a period of one or two years before the revanchist native vegetation full of weeds, resprouting shrubs, and soil microbes overwhelms the site and forces its abandonment. Agriculture in Mexico had seized an existing ecology of fire and redirected it to different purposes.

For a good six thousand years, smoke has filled the Mexican sky at the end of the dry season as swiddeners fire their milpas and *coamiles* prior to the onset of the rains. By controlling what is felled and what remains, agriculturalists gradually replace unusable species with useful ones. Ethnobotanists regularly record extraordinary relationships between such peoples and their surrounding landscapes. Globally, swiddeners exploit a very high fraction of the flora for some purpose or another, sometimes more than 90 percent of species present. One explanation is cleverness. Another is the extent to which their practices and choices have shaped that environment. Over millennia, as they returned to the same sites over and again every thirty years or so, they promoted the productive species and weeded out the worthless. Fire was the catalyst.

Agriculture has long been the core of Mexican existence. Mayan myths equate maize with humans. Today 95 percent of the maize grown in the Yucatán still derives from swidden cultivation. Even the Conquest did not remove its centrality, though Spain did introduce wheat to the mix, added pastoralism to farming, and shifted the state's economic obsession to precious minerals. Cattle, sheep, horses, burros, swine, oxen—all unknown prior to the Conquest—became increasingly central to land use. The regimes that resulted were different because animals and fire can interact with effects neither can yield by itself. They shape each other's habitat. In arid environments, intensive grazing can crop off the grassy fuels that feed fire. But elsewhere, where unpruned vegetation can overwhelm grasses, only fire keeps the land in pasture.

Weather, vegetation, people—these made the great fire triangle of Mexico. The resulting fire regimes, like climates, are a statistical composite. None of their elements cycles with mechanical rigor. Weather is chaotic; flora and fauna and people, living. Worse, humans respond to cultural considerations that place their fires in a moral universe outside the circuitry of ecological feedbacks. What has most organized the ensemble is agriculture. If fire made agriculture possible, agriculture in turn defined the terms of fire's regimes. Like the daily firing of the hearth, the seasonal smoke rising from milpa and pastizal testified to the fundamental rhythms of life. They are practices ancient beyond memory.

For that very reason, fire became increasingly invisible. Huehueteotl, the Nahuatl god of fire, was represented as an old man with a brazier on his head. He was Dios Viejo, the Old God. He was the core god associated with the four directions of the universe; his light and heat were echoed in the volcanoes, the stars, and above all the sun that framed the natural world. Less dramatically he was lodged in the rhythms of human life, in the home, in the fields, in the illumination of the everyday world; he was the protection against the night's terrors and shapeshifters. But as civilizations came and went, they added their tribal gods to the pantheon, eager to symbolize acquired political power with the triumph of their new myths and unique godheads. Throughout, however, the Old God remained, patient, enduring, much as fire persisted in the hearth and on the landscape while Tulas and Teotihuacáns rose and fell with an eruptive rhythm akin to the volcanoes on which they modeled their ceremonial pyramids.

History by cataclysm is a theme embedded in Mexico's mythology as well as in its landscape. Nahuatl cosmology imagined a succession of worlds created and destroyed. The Aztecs accepted four great collapses and renewals, one by each of the basic elements—earth, air, water, fire. They did not have far to go to find evidence. Tectonic plates converge on central Mexico, which rises like a squeezed wedge. Volcanoes sprout like bunchgrass; the earth trembles to the shuffle of earthquakes. Hurricanes and tropical storms slam into the land from both the Caribbean and the Pacific. Drought and deluge alternately bake and drown. And the place burns.

The basic pattern not only varies, but also, if two or more of its elements move beyond normal limits, becomes subject to episodic conflagrations. One disturbance cascades into another; one cataclysm leverages into a larger. It happens most spectacularly when hurricanes smash into tropical forests, slashing immense swaths through forests, which then dry and burn. After Hurricane Janet in 1955, 275,000 hectares burned in the Yucatán; after Gilbert in 1988, roughly 120,000 hectares burned. There is no reason to believe these events are modern innovations.

Something similar, less dramatic but more pervasive, happened in 1998. This year the parts of the normative fire triangle were individually more extreme, and their compounded consequences more eruptive. Sparks kindled green plants that normally would have smothered them. Little fires became big. Big fires exploded. A self-limiting mosaic became a monoculture of combustible biomass—gasoline with fiber. The El Niño–Southern Oscillation that governs gross Pacific climate, this year exceptional, created weather that ran off a century's scale.

First it was exceptionally wet, then exceptionally dry. The winter brought rains, climaxing when two hurricanes slammed into the Sierra del Sur in October, leaving in their wake mud slides, cholera, and rank combustibles. Then the monsoonal winds shifted. The normal southerly flow of moist weather off the Gulf of Mexico moved west and swept instead over the meseta, drying rather than wetting. Even cloud forests in Oaxaca and Chiapas could not extract moisture from the arid air. Then the northerlies that should have brought rain from the Pacific stalled. The worst drought in seventy years deepened and split the biota like cracking mud. Even perennial tropical forests began to shed their leaves, allowing more sun and wind to penetrate and dry. An intricate fresco of wet and dry fuels smeared into a great arid smudge. This year everything could burn. This year almost everything that grew became available fuel.

But droughts do not kindle by spontaneous combustion. The very absence of rain meant that nature's torch, lightning, was absent. Ignition belonged with people, who burned from necessity and carelessness, for arson and agriculture, from despair and calculation. Most of the season's fires, more than 80 percent nationally, started from escaped field burns. The campesinos had no choice: without fire they would have no crops. The problem was one of extraordinary conditions on the land and a demand for timing better suited to Wall Street than Oaxaca.

No one knew when the rains might return. Traditionally, burning obeys a seasonal rhythm, its frequency rising as the dry season concludes, the ash ideally warm to receive the rain. But that assumes a stability, a discipline, in nature and society that did not this year exist. The rains failed. The social order faltered. A great belt of fires along the Sierra Madre Occidental corresponded with sites ruled by *narcotraficantes*. The heaviest burning occurred in Chiapas, the scene not only of a dense flora but of open rebellion. Social unrest worked with, not against, environmental catastrophe. Everywhere, moreover, the campesinos burned, as they had to. This year in particular they resembled small investors playing a long bull market, unable and unwilling to get out. They had to stay in—had to burn—before it ended. Yet they were uncertain when nature's economy was likely to crash. This time it not only crashed but burned. As Nahuatl cosmology had always insisted, fire was a force to make and break worlds.

For this, too, there is precedent. Some sixty-five million years ago a great meteor smashed into the Yucatán, created the Chicxulub Crater, and probably contributed to the vast extinctions that define the border between the Cretaceous and the Tertiary. Almost everywhere, mixed with the iridium tracers that identify the event as extraterrestrial, charcoal crams the lithic horizon. Although the fireball likely accounts for some of this char, in the blown-apart forests it felled and in the cold dead forests and the stressed dying forests it left behind, fires long continued to enter and contributed to a world in which the reptiles would decline and the mammals and birds would begin to rule. The Chicxulub Crater of Doom has become the world's most popular image of a fire-catalyzed catastrophe. Yet buried along the crater is a force, a potential fire, more powerful than the meteor.

Beneath the overgrown surface, below the sloshing waters of the Gulf, lies an enormous quantity of fossil biomass, mostly stored as natural gas and petroleum. The earliest oil was found easily; drillers sank wells where the crude had already seeped to the surface. Exploration eventually revealed real gushers, however, world-class reserves were identified; and by 1921 Mexico was the third-largest petroleum producer on Earth. Oil replaced land as the major irritant in U.S.–Mexico relations. In 1938, Mexico nationalized foreign oil holdings. Petroleum now played the role silver once assumed for the Mexican economy. Oil became the bullion of an industrializing Mexico.

It could have been much more. Fossil biomass has made possible industrial fire, the most powerful fire agency since hominids claimed the torch as their own. Although the economy of oil is more or less understood, its ecology is not. Industrial fire has created a new biology of burning. Combustion powered by fossil biomass has the power to add and subtract fire and to rearrange fire regimes. From its inception in the Devonian, fire could thrive only as far as its fuels permitted. The triumph of anthropogenic fire lay not only in human control over ignition, but in humanity's ability to create new fuels from surrounding biomass. That, in essence, is what agriculture did. Even so, it was not possible to burn more than the available living fuel permitted, which meant that the ecology of a biomass-based fire was self-limiting. It is possible to reorganize matter but not to transcend the regime.

That changed when fossil biomass allowed for combustion to act on ancient fuels, no longer locked into the circular logic of what could be grown, harvested, and replanted. The sources of fire were, in practical terms, unlimited. So were many of fire's ecologic effects. Fire-distilled fossil biomass yielded fertilizers, herbicides, pesticides. Fire-driven machines could clear off land and chew up residues. Coal, oil, and gas were, in effect, fossil fallow, stored in the outlands of geologic time.

Where fire existed as a technology, industrial fire could replace it, and in most of the world where industrial and anthropogenic fire compete, it does just that. Industrial fire also has sustained the attempt to suppress open burning, even in wildlands. Without internal-combustion pumps, vehicles, aircraft, and tractors, fire control must resort to its primordial strategies, to protective burn before fire season and to counterburn against wildfires; both have the effect of keeping fire on the landscape. Still more widely, the reorganization of nature to support industrial economies has rearranged vegetation profoundly—which is to say, the fuel stratum of fire regimes. As evening satellite images show, the

Earth is dividing into two great fire moieties: one of free-burning fire, one of industrial combustion.

But not everywhere. The division has not happened, for example, in India, and it has not occurred in Mexico. Here the persistence of village economies, backed by bankrupt political rhetoric of colonial independence and revolution, has caused the two fires to compound. Industrial fire has plenty of problems. It has prompted nations to appreciate that fire is an ecological process, not simply a technological tool—that removing fire can be as environmentally powerful as applying it. Fire exclusion in wild or quasi-natural landscapes has ecological as well as economic costs. If industrial fire has solved, at least for several centuries, the problem of fire sources, it has substituted an equally implacable problem of fire sinks. The effluent from industrial combustion is too immense for old ecological sponges; the rate of output is greater than the rate of absorption; most goes into the atmosphere. Whether greenhouse gases will yield a true greenhouse effect is unclear, but the issue is one that could never have arisen within the ecology of biomass burning.

Thus, the fire industrialization of Mexico is unlike that of most nations. The two fires compound rather than replace. Industrial combustion assists but does not supersede traditional burning in communal *ejidos;* it augments the combustion load of cities without abolishing garbage fires, ditch burning, and wildfire through woody enclaves such as parks. Instead, both problems coexist. Patches of industrial fire coincide with or overlap patches of anthropogenic fire. Under extreme conditions, industrial fire ceases to behave according to normal ecological principles, indifferent to seasonality, incompetent to sustain biotas—an ugly atmospheric cocktail. There is too much of the wrong fire, too little of the right. The fires compete for the airshed and compound the biotic error.

Abysmal air quality is the most obvious arena for this competition. The Valley of Mexico, high and enclosed, is a special case, but it has served so long as the political and cultural center of the subcontinent that its symbolism matters as much as its lethal sky. Ozone, acid rain, pollutants—all weaken trees so that they are more susceptible to disease, fungi, and insects. The native woods, especially firs, are most damaged, and reforestation schemes replace them with exotics such as pines imported from other states and Australian eucalypts. Eventually the trees become fodder for free-burning fire. Instead of substituting for open burning, industrial combustion is creating more fuels for it.

The valley, with Mexico City at its core, is a megalopolis of perhaps twenty million people. Subsidized oil has made it possible, and industrial combustion makes it barely livable. The scene has come to epitomize what is wrong with Mexican industrialization—the social wreckage, the assault on human health and quality of life, the anthill crowding, even the corruption of politics. The great emblem of industrial nationalism, Pemex, has replaced the mines at San Luis Potosí as a symbol of an imbalanced, even dysfunctional economy.

In Mexico City, the Plaza of the Three Cultures advertises the synthesis of Mexican identity. The indigenes, the Europeans, the mestizos—their symbolic fusion represents the cosmic race of Mexico, even if they often remain segregated in practice. Not far away, in the mountains that frame the city's western perimeter, the Desierto de los Leones National Park reveals a comparable image for Mexico's fire history. The 1998 summer wildfires burned over forests scarred with lightning and sickened by air pollution. Along a mountain trail, it is possible to find a log gouged by lightning, the cambium infected with beetles and fungi, likely an epiphenomenon of industrial fire, the whole trunk burned to black cardboard. Call it the Place of the Three Fires.

The Cerro de la Estrella rises in the center of the Valley of Mexico. In pre-Columbian times, it was an island, where the waters of Lake Texcoco lapped with Lake Xochimilco. It was here, every fifty-two years, that the Aztecs celebrated the Ceremony of the New Fire. Then, when the two calendars, the 260-day and the 365-day, coincided, when the Pleiades stood overhead, when the cosmos was poised to crash into darkness or to rekindle into a new sun, the New Fire redeemed the world.

The ceremony was elaborate, its siting on the Cerro majestic. In all the countryside around, across the wide-mirrored lake, in every hearth and village, in every temple, in every torch and campsite, the fires were extinguished until all human light had vanished from the evening darkness. Only the illumination of the stars remained. The world—the known world of the sun—shuddered in uncertainty. The dark and the demons crept closer. Only a renewed fire, kindled in the ancient way, the way humans first learned to make it, could spark the sun's return.

On an altar at the Cerro's summit, four priests waited, one for each of the elements, for each of the four previous worlds, for each of the four thirteen-year counts whose beat summed up to a New Fire. A fifth priest ripped out the beating heart of a prisoner, the mandatory human sacrifice. A New Fire emerged from the sacred implements and was placed in the exposed chest to signify new life; then each of the four priests ignited a great torch from the common New Fire and, surrounded by guards, marched down the slopes to boats waiting to take them to each of the cardinal points. Ashore, the priests kindled the fuels of a subsidiary New Fire, each overseen by a priestess whose task it was to keep the fire burning for another fifty-two years. To fail was fatal. From this fire, all the fires of hearths, furnaces, and temples, all the fires used in hunting, farming, fishing, all the fires of life sacred and profane were rekindled. The stars would wheel in their places. The sun would rise. Once more, the world was saved.

The symbolic universe of the Aztecs ended with the Conquest. The New Fire Ceremony, last enacted in 1507, was never restored officially. Today little remains. Lake Texcoco is virtually drained. Houses fill the former lake bed. Smog replaces the lapping water. The effluent of industrial fire seals off the stars. Wildfire burns the hills. In macabre commentary, the fires of 1998 even swept over the Cerro de la Estrella.

The breadth of the 1998 season fires shocked officials and citizens alike. The burns were nearly everywhere visible, and nearly everywhere the popula-tion—foresters, politicians, citizens—believed they would lead to important reforms. They might. Big fires sometimes do. But not often, and they do not always elicit a suitable response. The political ecology of fire is subtle; fire's effects are very much contingent on what events interact with the fire; the ancient rhythms of Mexico's agriculture, beliefs, patterns of drought and deluge, and history by cataclysm are not changed easily. It is not obvious that the 1998 fires will serve as the New Fire of an industrializing Mexico.

The social ecology of fire is not unlike that of swidden. The opportunities for change are greatest in the first year, drop significantly after the second, and almost disappear by the third. Sites recover quickly; memory overgrows with the weeds of everyday life; the media moves feverishly on to new disturbances. Just as a fire's effects are prolonged and more profound if the site is grazed or logged or otherwise also manipulated, so a big fire can leverage its consequences if it connects to some other social, intellectual, or political movements. The degree of crisis matters. Too little shock, and the fires can be ignored safely as a climatic aberration. Too much—say, if the fires join a general wreckage of landscape and society, as in Indonesia in 1997 and 1998—and there may be little political will or institutional apparatus by which to deal with them.

Reform-minded Mexicans would do well to ponder these observations. They have a year, probably no more than two, to instigate official programs in response. That would coincide, morever, with the fundamental rhythm of Mexican politics: the six-year tenure of its powerful presidents. Traditionally every new occupant reorganizes and renames the bureaus under his direction. A fire program would have to survive that process and, in fact, survive the electoral collapse of the Partido Revolucionario Institucional (PRI) and the ascendancy of rival Vicente Fox to the presidency.

This time, however, the political context has altered, perhaps profoundly. Mexico is experiencing the greatest social turmoil since its revolution. It is moving away from one-party politics dominated by the PRI; it is feeling the economic impress of the North American Free-Trade Agreement; narcotraficantes in the north and rebellious peasants in the south are pushing the borders of internal politics, as immigration, legal and illegal, into the United States is redefining Mexico's relation with its superpower neighbor. The 1998 fires burned within an extraordinary context.

In searching for a suitable response, Mexicans have several levels of analogy open to them, many drafted from the United States. Mexico would be wise to scrutinize the Florida fires that flared as its own were dying out. Even the United States with its immense firefighting resources was helpless and had to evacuate almost one hundred thousand residents. Mexico has a long tradition of responding to internal unrest with military force; the Florida experience suggests the limits of that option with regard to fire. Although Mexico clearly needs to improve fire control, particularly at a community level, suppression will not solve the larger problem. Mexico undoubtedly also will look to American science and information technology. It will use the crisis to push for satellite imagery, computer modeling, radios, air tankers. Yet this, too, although necessary, may prove surprisingly limited unless it connects with a broader culture.

As Mexico looks for advice and examples, not only should it examine computer modeling of fire behavior and risk, explore the technology of global information systems, and assess the impact of the fires on biodiversity, it should develop its cultural consciousness of fire. It needs to find folk heroes among the campesinos killed fighting fire in Puebla; needs to perceive the incinerated hillsides around the Valley of Mexico as a threat to its national identity, as the smoke was to public health; needs to promote a response to the burned cloud forests at Chimalapas that is more culturally engaging than debris dams and reforestation; needs, finally, to find an Octavio Paz or Carlos Fuentes to incorporate the saga into its national literature. It needs, that is, to metamorphose the hundreds of thousands of burned hectares, the grim fatalities, the opaque smoke, the helpless and stinging embarrassment into a story that speaks to its deepest desires and sense of itself.

William James once summarized the pragmatic test of truth with a paraphrase from the New Testament. By their fruits ye shall know them, he announced, not by their roots. That captures a general American sentiment: the future matters more than the past; you are what you make of yourself. But Mexico still obsesses over its roots. A reformist fire program requires a link to the past. The obvious answer is fire, controlled fire, agricultural fire.

Unlike the scenario of the New Fire Ceremony, Mexico hasn't the luxury of extinguishing its panorama of fires and starting over. Much of Mexico remains under a regimen of agricultural burning. Not for a long time, if ever, will industrial fire truly substitute, and if no deliberate burning is done, nature or accident will impose fire in the form of cataclysms. Mexico must live with its cosmic race of mixed combustion. But the agricultural fire—better bred, less likely to become rabid, adapted to the needs of forestry and nature reserves—could well serve, probably must serve, as the basis for a New Fire. If Mexico can carry that vision away from the scorched slopes of the Cerro de la Estrella, it may well have found a formula for pyric renewal.

place
of
power

The United Nations Educational, Scientific, and Cultural Organization's (UNESCO) first symposium on environmental history, hosted by the Instituto de Ecología in Xalapa, Mexico, planned to conclude with a two-day field excursion to Los Tuxtlas, a cluster of volcanic peaks along the Gulf Coast in the state of Veracruz. Los Tuxtlas was a place of chronic unrest, fabulous biodiversity, ancient human occupation, and spectacular scenery—in brief, a place of power.

The climax occurred just before sunset, after a winding bus ride around the countryside and over a one-lane bridge through the town of Tres Zapotes. The bus shuddered to a halt on a dirt road. Cultivated crops spread in patch-fields to all sides; the volcano San Martín rose majestically on one flank and a dirt mound to the other. The mound—believed to be the site of an Olmec ruin, perhaps a small pyramid, though no one has excavated it to find out—swelled upward from a close-cropped pasture. The grasses and cattle flowed over the hill. Its summit was the object of our trek: here the themes of the conference could converge.

Our guides led us to the top, past munching cattle, and there one of the conference's Canadian hosts described how, in casing out this place as a possible tour site, he and the locals had joined hands as they crossed the field, bowed heads in assorted prayers at the summit, and sensed the place's aura. We were invited to share these sentiments, to sense the power emanating from the place by removing our shoes and touching the earth with our bare feet. It certainly testified to the power of political correctness: the genuflection to indigenous knowledge, the celebration of cultural and biological diversity, the need to reconcile modern science with moral if not spiritual powers.

But the "indigene" turned out to be a man who had moved to the region twenty-five years previously to carve pastures out of the rain forest. The mound was of indeterminate provenance. The cultivated landscape was an ecological mosaic, most of whose tiles had come from elsewhere—sugar cane from India by way of Europe, maize from the Mexican plateau, pasture grasses from Africa, cattle from Eurasia and Africa. The invitation to fuse directly with the dirt inspired, in my mind, a vision of Mexican ticks ripe with heaven-knows-what microbes fusing with my big toe. Then the whole ceremony was upstaged by a fire in an adjacent field.

All the way from Mexico City fires had lit our route. Ditch fires. Garbage fires. Patch fires along roads and rails. And especially field fires, for it was the season to harvest sugar cane, and the fields are fired prior to harvest and again to clean up the postharvest debris while the new cane shoots up (cane is a perennial grass; it just resprouts). So, as we stood on the mound, a campesino worked his way patiently around the plot, letting the flames push against the rising wind. The fire would burn away the dead leaves and tough husks and open the site for easier cutting. The smoke wafted up insects, and bird swarms darted among the puffs to pick them off. The heat flushed out lizards and snakes and other creatures, and more birds, like egrets, converged to hunt them. The wind gusted, and the flames blasted upward. However careful its choreography and tedious its translation, the lecture proved impossible to hear—burning cane is loud, like crumpled butcher paper. One by one, heads turned to the unplanned but predictable spectacle of the flaming field. The crackle of the cane spoke louder than theory. The real display of indigenous knowledge was the plodding campesino doing what you have to do to make a habitable place. The real lesson in power was the torch.

strange fire

the european encounter with fire

introduction: The Labors of Linnaeus

In 1749, Carl von Linné was at the height of his fame—professorially ensconced at Uppsala, popularly enshrined as a "second Adam" who had named all the plants and animals of the world, celebrated for his travels throughout Sweden, especially Lapland, renowned as the author of *Systema Naturae* and other classics of natural history written under his Latin nom de plume, Linnaeus. So it was natural that King Frederick I should invite him to tour Sweden's most southerly lands, Skåne, that Linnaeus should agree, and that he use the occasion to return through Småland, his home province.

There he studied the contentious subject of *svedjebruk*, Swedish swidden, "looked upon by some as profitable, by others as rather deleterious." So prevalent was the practice that some Linnaean contemporaries (wrongly) believed Sweden had derived its name from the endless *svedje* that comprised its countryside. Pondering the evidence, Linnaeus concluded that, although fire unquestionably consumed humus ("the food of all growth"), the practice allowed farmers to get "an abundance of grain ['and for several years . . . a good pasture of grass'] from otherwise quite worthless land." Deny them that burning, and "they would want for bread and be left with an empty stomach looking at a sterile waste . . . a thankless soil and stony Arabia infelix."

The passage, however, outraged Baron Hårleman, high commissioner of agriculture, and Linnaeus was forced to delete it from the final publication. "Not only," Hårleman fumed, had Linnaeus "not condemned svedjebruk, so pernicious for the country, but even contrary to his own better judgment justified and sanctioned the undertaking." In penance, Linnaeus was compelled to insert a long passage on the value of livestock manure as a way to supplement traditional forest composts of heather, moss, and conifer needles. Even in High Enlightenment Europe, it seems that burning could not compete with BS.

As a dialectic between humans and nature, fire regimes express the values, institutions, and beliefs of their sustaining societies. The labors of Linnaeus tell us a great deal about European fire, about the extent to which it was embedded in the social matrix of agriculture, and hence why it looks invisible to those, such as North Americans and Australians, to whom fire is recognizable only when it free-burns in wildlands and bush. But the episode also alerts us to the peculiar problems that would arise when a civilization such as Europe's would encounter lands far more fire prone than its sodden core and would deal with folks far more fire addicted than European intellectuals. Compared with Aboriginal burning in Australia, agricultural and pastoral fires in Africa and India, landscape-sweeping burns across North America, and colonizing fires in new settlements everywhere, svedjebruk was as quaint as a Midsummer's Day bonfire. So, too, the outrage provoked by rural burning in tiny Skåne would pale before the dialogue inspired by continental-scaled firing that would confront an imperial Europe.

Those differences matter. Fire on Earth looks the way it does today because Europe expanded beyond its constricting peninsulas and islands to become a global power and because it vectored to that overseas imperium an industrial revolution that could, by exploiting fossil hydrocarbons, transcend the endless ecological cycles of its agricultural heritage. Europe has influenced fire in Asia, Africa, Australasia, and the Americas in ways that none of those places have influenced European fire. From Europe's ancient association of fire and agriculture came conceptions of fire's proper place in the landscape. From Europe's alliance of forestry and imperialism came the attempt to suppress fire in large forest and wildlife reserves created in overseas colonies. From Europe's industrialization came the apparatus to enforce the agenda of fire abolitionists. It is no accident that the continued condemnation of fire by international environmentalism—from nuclear winter to greenhouse summer, from fire as emblem of social disorder to fire as a perverter of biodiversity—has its origins in Europe. Europe's fire has become, as Europe always believed it would be, a vestal fire for the planet.

i.   An Endless Cycle: Fire and Fallow

Before there was a Europe, there was fire. But as the last of the Pleistocene glaciations receded, that fire was anthropogenic, and increasingly it was prescribed by the imperatives of the Neolithic revolution. The hominid colonization of Holocene Europe was a flame-catalyzed reclamation by agriculture. Over long millennia, livestock and cultigens, tended by fire-wielding humans, penetrated into every valley, propagated up every slope, prevailed over every alternative biota. The social order dictated the biotic order. Europe became an immense garden.

There was no lack of fire. In mediterranean and boreal Europe, agriculture simply replaced one fire regime with another. In central Europe—the temperate core of the continent—the Neolithic vigorously expanded the dominion of fire. Here agricultural colonizers shattered the biotic bell jar of shade-lording lindens and elms and concocted a biotic stew that simmered over a succession of landscape-cooking fires. Fire cleared new fields; fire renewed old ones; fire flushed pastures; fire stimulated forbs and berries and pharmaceuticals suitable for foraging. Fire established the garden; fire cleaned it periodically; fire powered the dynamos of its nutrient cycles. Without fire, much of Europe was uninhabitable. Practitioners never doubted this fact.

Europe was a dense miniature of landscapes and peoples, and the geography of its fire reflected this complexity. Yet all Europe's agricultural systems, excepting (with qualifications) irrigated croplands, were variants of fire-fallow farming practices. Some land lay fallow for a season, some for one year out of a handful, some for decades while cropland succumbed to rough pasture and then to woods. There is no good name for this collectivity of fire-rekindled fallowing. *Swidden* is the common anthropological expression, but when it was introduced to the literature to describe East Asian practices in the 1950s, it was greeted with disdain (and dismay). Critics such as H. H. Bartlett preferred indigenous terms, of which there were dozens. Why, he asked, accept for the Philippines and the East Indies an antiquated expression drawn from Viking-colonized Yorkshire to describe the burning of the ling? In 1958, the word remained so obscure that it was not even recorded in the *Oxford English Dictionary*. But the term triumphed, anyway, a metaphor for the ascendancy of European concepts throughout the globe. It seems only appropriate to have it apply, with all its faults, to Europe itself.

Its fallows, like its poor, were always among Europe's farmlands. Intellectuals hated such places. With few exceptions—Virgil, Linnaeus, a handful of others who had grown up on farms—professional agronomists, all trained in cities and housed in academies and bureaus, detested the fallows as a waste of productive land and an invitation to sloth. Worse, the fallows were burned. For a civilization constantly pushed to (and beyond) its demographic limits, haunted by visions of famine and hunger-driven disease, the flaming fallows were unconscionable. The "disgrace of the fallows," as François de Neufchateau declared in the eighteenth century, obsessed agricultural intelligentsia and officials. Why did they persist? Why could they not be utilized instead of being sacrificed to the faithless flames? Why could not cultivation proceed to the point that it dispensed with fire altogether?

There was no single reason. But agronomists had the relationship reversed. Fire did not follow fallow as plague did rats. The burning was not a convenient (if, to official eyes, indolent and reckless) means to dispose of agricultural trash. The fallows were grown to be burned. They were cultivated to support fire just as three-field rotations grew oats and barley to feed draft animals. Fire was integral, not incidental. Agriculturalists relied on plants to recapture nutrients and then burned them to liberate those biochemicals in a suitable form. They needed to jolt fields back to life, needed to purge soils temporarily of hostile microorganisms and weeds, needed to flush stale pasture with succulent proteins. Fire alone did this, and, in order to burn, fire required suitable fuels; these were grown. Even constantly cultivated infields required outfields from which to gather combustibles or as places to run the herds from which, housed in winter barns, manure could accumulate. Those outfields were swiddened and their rough pastures burned.

Fire and fallow constituted an endless cycle—now swelling outward, now contracting, but never broken—that informed European agriculture and through it shaped the European landscape and informed a European land ethic. Where officials saw environmental and social wastage—lost revenues, wandering swiddeners and pastoralists, incinerated soils, scorched timber—peasants saw renewal, and their fire ceremonies clearly spoke to the dual virtues of fire to purge the bad and promote the good. Fire was as essential to the farm as to the household, a tool more indispensable than plows and harrows. By contrast, an urban intelligentsia experienced fire in cities and identified it with social disorder. Intellectuals denounced; peasants burned.

The official protest did have its logic. However much agriculture reshaped landscapes, it ultimately could not make something out of nothing. The prevailing understanding of nature was that it was profoundly cyclic. Growth and decay were exquisitely balanced; since the Creation, no species had been added to or subtracted from the Great Chain of Being; agriculture required that no more could be removed from a field than had been previously put into it. Productivity could increase only by building up the reservoir of soil nutrients. Failure to return as much as was harvested led to degradation, of which there were endless examples. In this Newtonian ecology, landscapes orbited with a biotic balance as delicate as that the planets traced between gravitational and centrifugal forces.

Each group perceived the central fire differently. What farmers saw as a biotic hearth, agronomists saw as an ecological auto-da-fé that burned away the carpets and walls. Agricultural fire condemned civilization to a rural rut and prevented any hope of improvement. That fire burned away humus was thus sufficient cause to criticize it; that it promoted wasteland and fallow was enough to condemn it. Fire took away: it did not give back. The transmuting fire was, to critics, a kind of folk alchemy, proposing to turn environmental lead into agricultural gold. Folk fire practices seemed no better than superstition, a ritual with no more substance than ancient fire ceremonies that burned witches and heretics. For millennia, fire traced the great divide between "rational" and "primitive" agriculture.

Nature's economy was inseparable from society's. It did not help fire's cause that within a landscape such as Europe's, shaped by human artifice, wildfires appeared most prominently during times of social breakdown. Anything that left the garden untended—war, revolution, famine, pestilence—would let fallow run riot and would end, like peasant rebellions, in torch and sword. Both society and land were bound by the same inexorable logic, and fire threatened each.

Even the lordly Linnaeus could not escape, nor wished to. The same year that he toured Skåne, he published his widely influential essay *Oeconomia Naturae* (The Economy of Nature), in which he conveyed the exquisite checks and balances that informed nature's polity and that equally governed cultivation. The theme obsessed him. In his last, brooding essay, he pondered the character of divine justice by which every deed had, with almost Newtonian logic, its retribution.

But there was no doubting fire's presence and power. When Linnaeus was still a youth, Herman Boerhaave had declared that "if you make a mistake in your exposition of the Nature of Fire, your error will spread to all branches of physics, and this is because, in all natural production, Fire . . . is always the chief agent." Around that central fire, Europe's agricultural systems orbited. Even long hunters in Finnmark, transhumant pastoralists in the Apennines and Pyrenees, and Slavic swiddeners in Siberia were caught in its ecological force field like swarms of biotic comets.

ii.   Extending the Cycle: Enlightenment and Expansion

But if Linnaeus could not escape, his Apostles—the twelve students he dispatched around the world—could. Symbolically they extended the circle of the European Enlightenment and with it a dynamic of exploration and empire that established a European hegemony not only in the world's political economy but in its scholarship. That propagating periphery was all too often a frontier of fire.

Europe reconstructed how it thought about the world, and it began rebuilding the world accordingly. Intellectual Europe increasingly accepted modern science as the model of knowledge, and it enshrined progress, not renaissance, as nature's informing principle. Imperial Europe renewed rivalries on a global scale in what William Goetzmann has termed the Second Great Age of Discovery; overseas outposts moved beyond trading factories on the coast and probed boldly inland, repopulating landscapes with European émigrés, remaking foreign lands into a constellation of Neo-Europes, a colonial outfield to the metropolitan infield. Industrial Europe, building on a century of agricultural reformation, experimented with an economy no longer circumscribed by humus, manure, and sunlight.

Carrying the Torch: New Lands, New Fires

The Maoris, the Malagasy, the Madeirans—all have founding myths of a great fire that accompanied settlement, and the record of such fires is buried in their soils and lakes. Similar contact fires have left comparable records in Brazil, Iceland, and Australia—everywhere that had colonizing peoples and the means to preserve charcoal. That imperial Europe also should have its world-transmuting fires is no surprise. Everywhere, the strike of European steel on indigenous flint threw sparks to all sides. Forested frontiers, in particular, were a flaming front full of eruptive fires with names such as Peshtigo and Miramichi and Black Thursday that eventually left behind a landscape of more subdued, residual combustion.

And fire *did* remain, sometimes a cause of these immense changes, often a consequence, always a catalyst. Europe's expansion brought it to lands notoriously susceptible to fire—environments for which seasonality meant a movement between wet and dry, not cold and hot; biotas salted with pyrophytes; landscapes already baked in a hominid hearth. In places, colonization expanded fire by transferring European fire practices to receptive landscapes or by breaking down old biotas and reconstituting them with Eurasian surrogates. Thus it was that swidden exploded out of New Sweden across the American backwoods frontier; that sheep swirled around the gyre of Australian grasslands; that livestock crushed the once-grassy deserts and savannas of the American West and the South African karoo. The exotic fauna soon spread exotic flora from wild artichokes to cheat grass to tumbleweeds, and fires rose and fell with the strength of these biotic tides. Elsewhere, colonization sought to contain indigenous fires or to abolish them.

Settlement ultimately exchanged one fire regime for another. Colonizers, in the end, could not deny the logic of an Earth fluffed with combustibles and marinated in oxygen. There might be more fire or less fire; there might be efforts to suppress certain kinds of undesirable fires kindled by natives, malcontents, and ignorant immigrants; but there was no expectation that fire itself would disappear or that fire practices could vanish like old smoke. Fire would simply be glossed into the text of the new landscape, the rubrics of a flame-illuminated manuscript.

Certainly this was true for North America and Australasia. The redefinition of fire regimes is exactly what C. S. Sargent's map of U.S. forest fires for the 1880 census, Franklin Hough's 1882 *Report on Forestry*, and legions of settler journals corroborate. America, Canada, and Australasia were developing landscapes, most of whose peoples lived by agriculture. Conflagrations might rage during the time of transition, but great wildfires required wildlands, and they would cease as settlement became sedentary and as controlled burning found its niche in the cultivated countryside. Meanwhile, there was little doubt that fire was an inextricable presence and proper fire use the best means of fire control.

Fire Conservancy: Protecting Land, Preventing Fire

Still, Europe's colonies presented some novel circumstances. There was a natural desire to control the worst fires rather than to rely on their laissez-faire recession; and there was general recognition that in new worlds opportunities for wholesale reform were possible on a scale never before imagined. In its most dramatic expression, colonizing powers began to reserve lands, typically forested, for the public good. These reserves were an imperial invention; they were possible because the lands were (or partially could be made to be) vacant; and they created an imperative for institutions to administer them. Especially in Asian and African colonies, fire joined famine, malaria, and banditry as a perceived blight on lands mired in fatalism and rural inertia—colonies for which railroads, telegraphs, civil service bureaus, and fire protection could promise hope, progress, and humus-laden soils. What the British called "fire conservancy" had analogues everywhere the imperium of the European Enlightenment spread.

Had those landscapes simply disappeared into a reconstituted agricultural regime, the Earth's combustion calculus would little resemble today's. The geography of free-burning fire among the industrial nations traces precisely the creation of official wildlands. It is worth noting how extraordinary these

institutions are, how exceptional their presence in world history, how fragile
their survival. The oldest do not date beyond the 1860s. Most have a tenure
of less than a century. New Zealand already has begun their disestablishment.
Reversions, some large, are promised to indigenous peoples in Canada and
Australia.

The wholesale reservation of lands for the public good was an obvious if
idealistic solution to human misuse. Remove humans and you remove abusive
human practices. (There was little sense that much of what was attractive
about colonial landscapes had resulted from long human manipulation.) The
particular arguments for such reserves were that they would ameliorate the
climate and decouple the ecological syllogisms by which deforestation seemingly led to desiccation. On previously uninhabited islands such as St. Helena,
Tobago, and Mauritius, Europeans immediately introduced fire, and as the
vegetable mold went up in smoke, so, it seemed, the local climate became
droughty, springs dried up, rare flora and fauna perished, and once-Edenic isles
degraded into cinders. Those microcosmic experiments foreshadowed macro-
cosmic doom as similar scenarios appeared to be unfolding before the eyes
of European savants wherever colonization, like a steel wedge, had cracked
open the continents. The most direct way to intervene was to deny access to
pioneering peoples.

What made such reservations politically plausible was that the land was
more or less emptied of humans. Those landscapes that were largely vacated
at the time of settlement—North America, Australasia, and, in a different
sense, Siberia—became the particular landscapes for public reserves. Disease
mostly had done the task; where it failed, relocations, colonial and internecine
wars, and overall social disruptions had pruned populations and prevented a
demographic recovery. But because of this population collapse, for a period of
time, the anomalous had become the norm: people were gone or rapidly going.
In North America and Australia, the great era of land reservation occurred
precisely at this instant, when the indigenes were fading and the colonizing
Europeans had not yet arrived in great numbers. A few decades either way and
it is unlikely that those reserves could have been made. They were a magnificent
historical accident.

One important consequence was that, to exploring Enlightenment philosophes, these landscapes seemed preternaturally wild. Relic bands of indigenes appeared no more competent to shape the scene than did those minor streams that occupied the great valleys previously scoured by Pleistocene glaciers. The land seemed fresh from the Creation. Natives dissolved into it like salt in the ocean. In reality, this perception was a freak of historical timing. The reality was that the land had been occupied as fully as peoples' technologies had allowed, that most landscapes were shaped intensively by indigenous practices, that many landscapes in the Americas and Australia were as fully anthropogenic as any found in Europe, that much of the New World had experienced human settlement far longer than had the Old. But during contact the land had gone feral. Colonizing Europeans, critics complained, had found a garden and left a wilderness.

Where the indigenous peoples persisted, the reserves were always compromised. Even a handful of graziers could wield enormous power. Here, reserves endured because of political force, backed by legal statute or military cantonment. Such quasi-inhabited landscapes continued to burn. This practice was as true for southern crackers in Florida and backblockers in Gippsland, Australia, as for Ghond tribesmen in central India. Whatever the legal regulations regarding access, without suitable fire the reserves were biologically inaccessible, so the locals burned. Moreover, fires could ecologically void edicts, fences, and patrols. Once the natives recognized the value officials placed on fire control, they had a ready weapon of protest. Woods arson became endemic.

But where the reserves remained essentially uninhabited, there was no obvious problem with anthropogenic fire. Nor was there an imperative to install a new system of controlled burning. Agricultural fire was necessary only if one lived by farming, grazing, hunting, or foraging on the land. The administrators of the great reserves did not: they only protected those landscapes. They accordingly were led to an anomaly as great as the vacant lands themselves. They could see fire exclusion as not only admirable but possible.

Forestry and Fire: The Paradox of Fire Protection

That foresters, not *jägmeisters* or agronomists or civil engineers, inherited the task of administration had enormous repercussions for the reserves. Colonial foresters saw fire through a peculiar prism and almost always found themselves fiercely at odds with indigenous practices, all of which exploited fire. It is no accident that the first question asked at the first symposium on the first large-scale experiment in forest reserves—the 1871 forest conference in India—was the necessity and practicality of fire control. "There is no possible doubt," wrote Lt. Colonel G. F. Pearson, conservator of forests for the Central Provinces, that "the prevention of these forest fires is the very essence and root of all measures of forest conservancy." Foresters everywhere echoed that sentiment.

This paradoxically was not the case in Europe itself, whose woods continued to simmer over chronic fires. Silviculture, after all, was a graft on the great rootstock of European agriculture, and modern forestry was part and parcel of the Agricultural Revolution that had preceded the better-known Industrial. Agroforestry was the norm, not a novelty; the fire practices of agrarian Europe were those also typical of its woods. Even in 1870, as much as 70 percent of the Schwarzwald was subject to a swidden cycle that involved cereals, root crops, and oak for which fire was fundamental. At the end of the nineteenth century, French forests around Provence regularly practiced *petit feu,* in which strips were burned protectively in regular rotation. Controlled burning persisted in the Ardennes forests until World War I, in the woodlands of the Midi until the 1920s, and in the Baltic pineries, rich with heath, until the Second World War. Fire was as prevalent and as essential in Europe's forests as in its fields.

But if woods workers exploited controlled fire, forest overseers hated and feared it. The political dialectics that informed agriculture—the divide between theorists and practitioners, field officers and government ministers, periphery and metropole—also shaped forestry. The brief against fire was that it destroyed trees better suited for timber or charcoal than for field ash, that it encouraged graziers and swiddeners to encroach on protected woods, that it eliminated the humus that was the universal index of ecological health. In nature's economy, fire was a prodigal heir who spent his nutrient capital rather than living off his annual interest. Following fire, biotas degraded; soils eroded; weeds proliferated; hillsides slid; torrents rushed; climates degenerated; and societies became unstable. The glory that was Greece, the grandeur that was Rome, the mystery that was the Maya—all succumbed to the internal rot of deforestation by that unholy alliance of cutting, grazing, and burning. To Europe's laborers, fire was a good servant but a bad master. To Europe's rulers, fire made for rich parents but poor children. For centuries, however, the only practical means of fire control was close cultivation and prescriptive fire use.

This was no less true in nineteenth-century settler societies. The best hope was to change fire regimes as painlessly as possible, not to eliminate fire altogether. Study Sargent's 1880 fire map of the United States, and you see a developing nation, profoundly agricultural, remaking itself with steam and industrial capital, but one for which any talk of fire exclusion was utopian nonsense. The character of fire had changed and would change further. Although no one argued that conflagrations like those that swept Gilded Age Wisconsin and Michigan were desirable and no one wanted to repeat the 1851 holocaust that seared Victoria, no sane critic argued that fire itself could be expunged. When in 1898 Gifford Pinchot thundered that "the question of forest fires, like the question of slavery, can be postponed, at enormous cost in the end, but sooner or later it must be faced," no one took his abolitionist analogy literally, not even in Gippsland, then suffering through Red Tuesday. The problem was to subdue wildfire, and the solution, as always, was to tend the garden and burn the fallow carefully.

Counterfire: Last Fires, Lost Fires

Yet imperial Europe pursued a radically different fire ambition in its colonies than in its own cultural hearth. Ideals long confined to hypothetical islands and practices safely caged in ancient social contexts were transported across the seas and released, propagating like the rabbits on Porto Santo or incongruously imposed like utopian colonies planted in Paraguay. Why? One explanation is political. European powers were prepared to behave differently toward colonial peoples than to their own. European forces could control indigenes in Cochin, Natal, the Maghreb, and New South Wales in ways that they could not control peasants in Galicia, Dalarna, and Provence. By the mid–nineteenth century, forestry had become, in fact, an inextricable part of European imperialism.

Colonial forestry was a composite, a kind of intellectual and institutional plywood that glued together the separate veneers of Germanic silviculture, French *dirigisme*, and British imperialism. The Germans were supreme as silviculturalists; the French had welded forestry to the purpose (and power) of the state; and the British—who at the start of the nineteenth century had lacked both forests and foresters—had fused the two into a package suitable for export. For Greater India, it hired German forest conservators, trained students at the French school at Nancy, and shipped British cadets throughout the empire. (When that imperium collapsed, so, almost overnight, did British forestry.) Until then they promulgated a transnational culture of fire control, carried equally to Cyprus, Sierra Leone, Tasmania, Cape Colony, New Zealand, even Hong Kong. It is instructive to recall that Americans such as Gifford Pinchot, Henry Graves, and Theodore Woolsey Jr. passed through this same curriculum; that Pinchot corresponded at length with Sir Dietrich Brandis, doyen of the Indian Imperial Forestry Service (and honored as the "Father of American Forestry"), about how to establish such an organization; that Pinchot later remarked that American foresters had much in common with French colonial foresters in Algeria (!).

This unfortunate comment should rattle our consciousness and remind us how much the public lands and public forestry were an imperial invention. Even in Europe, foresters existed primarily as state officers, either to promote wholesale reclamation of wastelands such as the Landes, the Hautes-Alpes, or the Jutland heaths or to intervene between local economies and the larger commercial forces against which traditional practices had proved incompetent too often. After the French Revolution had abolished the old forest regulations, an orgy of cutting and burning had prompted an environmental Reign of Terror that had left many communal forests in ruins and, more ominously, had stripped alpine slopes to the point that torrents propagated like swarming locusts. State-sponsored reforestation became the preferred remediation, further consolidating the confederation between forestry and the nation-state. That alliance was strengthened overseas. Prominent foresters behaved like proconsuls, moved in the highest circles of imperial administration, and were knighted for their service to the empire. Forestry behaved as an enlightened despot for the environment.

It is instructive that major imperial powers of Enlightenment Europe were from Europe's temperate core, from lands that did not experience routine fire seasons, that understood fire as an artifact of agriculture. The great powers were also the dominant industrial nations and scientific authorities, and they dismissed indigenous knowledge as shamanism and folk practices as ecological acupuncture. For colonial landscapes that had known only traditional practices, European forestry was a revolutionary force. For colonial authorities, it was a means of wholesale biotic rationalization and social reform. Through forestry, the colonizers would transform irregular wildlands into regularized woodlands, just as institutions of European jurisprudence would transform the jungles of folk legalisms and as railroads would restructure subsistence economies. Fire control was as fundamental to colonial rule as military garrisons, plantations, and acclimatization societies.

All this, however, is still an argument to seize the torch, not to extinguish it. It does not explain why foresters believed they should seek to abolish fire. One reason is that they could envision fire only as anthropogenic. Everywhere they looked, human burning had overwhelmed and defined the landscape; controlling fires and controlling people were one and the same task. Where reserves successfully excluded humans, however, it seemed possible that fires, formerly tolerated as a necessary evil, likewise could be banished. Only later did the potential for lightning fires become problematic. (In a sense, by removing the domesticated fire, foresters would allow the feral fire to replace it.) Fire exclusion ironically became plausible only because of the exclusion of people. The exemplar of uninhabited isles such as tropical Mauritius proved less an ideal than an aberration, a fata morgana of agricultural philosophes, and so, too, fire control untrammeled by normal politics—call it imperial Europe's suppression paradigm—shimmered over colonial forests like blue haze.

A second reason follows: foresters themselves did not live in and off those reserves. They protected land; they did not cultivate it. Had they been forced to inhabit those landscapes, they would have been compelled to manipulate them with fire, as people everywhere did, as in fact their professional brethren in central Europe did. They would have had no choice. But they were guards, not gardeners, so it seemed possible—given sufficient political will—to do in colonies what was only quixotic in Europe: to remove fire entirely from the garden. In those colonies densely populated with indigenes, the scheme failed. In those lands effectively depopulated, the experiment could proceed for several decades before its full ecological and economic costs became overwhelming.

Thus, forestry behaved differently outside its originating lands. Like so many other European utopias, forestry's necessarily was situated across the

Western Sea. The suppression paradigm, released from its originating social ecology, spread like blackberries in New Zealand or cheat grass in the Great Basin. In Europe proper, the quarrel over fire practices was ancient, and the balance of land-use power prevented a massive extinction of burning. What finally snuffed out those flames was the spread of industrialism, the fossil fallow of coal. The Schwarzwald swidden, for example, expired when steam transport rendered its oak-derived tannic acid no longer competitive against South American imports.

In Europe's colonies, the confrontation over fire practices was starker and shorn of traditional checks and balances. Some form of fire protection clearly was mandatory, but what form it should take was not obvious. The collision of European forestry with indigenous landscapes everywhere sparked a public debate about appropriate fire practices and policies. The celebrated controversy over light burning in the United States had cognates in the debates about early burning that kindled throughout the British Empire. All ended with official condemnation of burning, with the political clout to attempt to enforce that edict even beyond the reserves themselves, and, after forty to sixty years, with a recantation.

The reasons for that universal failure are all too familiar. The world's biotas obeyed rhythms different from those of temperate Europe; especially where landscapes experienced strong gradients dividing wet and dry seasons, fire persisted in defiance of agronomic and political theory, and ecosystems displayed fundamental adaptations to it. Fire exclusion rendered biotas less stable, less useful, less diverse, and less amenable to fire control. For a while, fire protection could be made to work. For a period of time, installing a first-order infrastructure for fire prevention by eliminating indigenous fires, by suppressing long-smoldering fires, and by actual firefighting could reduce burned acreage dramatically. But that grace period would not last. Either the land was converted to some other, less-flammable condition or else some species of controlled burning had to be introduced; otherwise, nature's economy would undergo violent cycles of fire booms and fire busts.

The mistake everywhere was not that Europeans sought to impose a new fire regime or that they fought wildfires to that end, but that they sought the abolition of fire. They failed to recognize that the removal of fire was as powerful

an ecological act as the introduction of fire. They believed that suppression
of fire would liberate oppressed biotas, much as the suppression of famine,
typhoid, and Thuggee could free backward societies to progress. Over and again
they interpreted fire in political rather than in environmental terms, as the
graffiti of ecological vandals, as the torches of barn burners and rural vigilantes,
as the protest of a folk both sullen and prescientific. The tragedy in America,
in particular, was not that wildfires were suppressed but that controlled fires
were no longer set.

Even so, early fire control necessarily relied on fire use, as forest guards
early-burned strips around reserves and underburned and then fought large
conflagrations with equally large backfires. Until some alternative pyrotechnol-
ogy appeared, fire exclusion was a concept as metaphysical as the geography
of Gog and Magog or Thomistic arguments for the existence of angels. Until
then, paradoxically, the imperial Europe that had sought to abolish free-burning
fire had created, with its immense colonial reserves, precisely the ideal circumstances for the perpetuation of fire.

iii. Transcending the Cycle: Industrial Combustion and Fossil Fallow

When Carl von Linné died in 1778, the Linnaean landscape of fallow and field,
with its ecology of closed cycles and the unbroken Great Chain of Being, was
beginning a fast disintegration. Two years earlier Captain James Cook had sailed
on his last voyage, confirming an immense age of European exploration and
empire; Thomas Jefferson had written the Declaration of Independence, not
only inaugurating an era of democratic revolution but announcing a colonial
breakout from the coast into the interior, a vanguard of Neo-Europes populated
by emigrant settlers; Adam Smith had published *The Wealth of Nations*, proposing a new economic order; and James Watt and Matthew Boulton had consoli-
dated their partnership to manufacture steam engines. The closed loops of the
Linnaean world fractured; expansion metamorphosed into evolution; nature's
cycles became spirals; and society's orbits became the time lines of progress.

Of these momentous events, the steam engine was, for fire history, the
most profound. Its demand for fuel soon exhausted biomass reserves and
compelled the custodians of industrial combustion to exhume lithic biomass

from the geologic past. First coal then petroleum and gas were a kind of biotic bullion that acted on nature's economy like the plundered wealth of the Aztecs and Incas did on imperial Spain's. No longer was combustion limited by the self-regulating ecology of grown fuels. Fossil fallows could replace living ones; steam transport could restructure the flow of nutrients so that ecosystems aligned with the routes of commerce and the trophic flow of capital; novel pathways of energy could be built on the controlled combustion of fossil fuels. Industrial pyrotechnologies could replace the fire practices of open burning. It was possible to transcend the closing circles of ecology. It was even possible, it seemed, to apply the new combustion not only in exchange for the old, but to suppress it.

This was critical. However fanatical foresters might be regarding free-burning fire, they could only change, not destroy, fire if they wanted to retain forest wildlands. Fire *protection* meant redesigning fire regimes, not abolishing them; fire *control* required burned fuelbreaks, preemptive burning around pro-tected sites, and regular *petit feu*; *firefighting* meant burning out and backfires. In the end, fire remained. In India, imperial foresters discovered to their dismay that, in order to secure prime groves, native guards were early-burning all the surrounding forests. Fire exclusion was possible, at best, for only a few years. European forestry could no more escape fire than could European agriculture; forestry was a long-fallow swidden, growing oak, pine, and *sal* instead of corn, barley, and cotton. Eventually fire would return. Eventually, that is, until industrial combustion arrived.

With power pumps, fire engines, aircraft, tractors, power saws, and motor-ized transport, it became possible to move firefighters into the reserved lands quickly enough to catch fires while they were still small and to meet free-burning fire at its early stages with a counterfire force of equal magnitude. It was possible to impose a relative condition of fire exclusion for a longer duration. This condition cost money; it demanded ever larger investments of technology and firefighters, and it wrenched the biota into successively greater distortions, but it could be done. In the short range and with state sponsorship, for which costs were not matched against the values protected, suppression by rapid detection and initial attack was the most effective means of fire protection.

Area burned from wildland fire shriveled. For settler societies, America replaced Europe as oracle and exemplar of fire protection.

In fact, fire receded everywhere, telegenic conflagrations notwithstanding. Fires disappeared from the domestic economy, from industrial pyrotechnologies, from agriculture, from forestry, and from many wildlands. Wherever fire was a *tool*, it had to compete with the new pyrotechnologies and generally lost. Whatever its ecological merits, open burning existed within an anthropogenic landscape for which free-burning fire and flame's obnoxious side effects such as smoke were no longer desired or even tolerated. The free-fall of free-burning fire promised to stabilize only where fire was essential as an ecological *process* for which no surrogates were possible. Like grizzly bears and Karner blue butterflies, fire retreated into special preserves. The geography of fire became one of massive maldistribution—too much wildfire, too little controlled burning; too much combustion, too little fire. By the 1990s, U.S. public lands were immersed in a crisis of "forest health" that was provoked, in good measure, by a fire famine.

Of course, the ideology of fire control mattered. Without the vision of fire exclusion, there would be in Europe's settler societies today more anthropogenic fire, less ecological havoc, and a better equilibrium between fire use and fire control. But the larger trajectory of fire history argues that industrial pyrotechnologies were destined to substitute wherever possible and that free-burning fire would recede into those habitats deliberately reserved from human habitation. The most profound of the modern epoch of extinctions may be that of anthropogenic fire. The epic of settler society fire, like all epics, is a tragedy.

Conclusion: Strange Fires, Still

Linnaeus could never have imagined a world without fire. On his travels, he met wildfire in Lapland, and throughout cultivated Sweden he encountered controlled fires for farming, forestry, and grazing; for the production of tar, potash, lime, and charcoal; and for every other imaginable human endeavor. No less than its woods, the age structure of Sweden's towns also betrayed their fire history. The hearth shaped the house; the dynamics of house fires shaped cities. Some landscapes needed more fire, some less, and others a different regimen. But fire there would be. The abolition of fire was as hypothetical as the extinction of water or dirt.

What sort of fire was acceptable? In the Old Testament, Leviticus tells the story of Nadab and Abihu, who place a "strange fire" before the altar of the tabernacle. Proper offerings the Lord accepted by a consuming flame. This strange, profaning fire, however, met with a "devouring" fire that slew the two sons of Aaron for their sacrilege. So, likewise, Europe's intellectuals and authorities knew what fires belonged on the land's altar and were prepared to reject profane substitutes with violence if necessary. The proper fires were those prescribed by the overseers of Europe's social order.

Assuring the proper fire mattered not only because unsuitable flames threatened property (and lives), but because they challenged a conception of society whose environmental expression was the garden. Too often fire encouraged movement. Swiddeners, pastoralists, long hunters, trappers, settlers—all exploited fire as part of seasonal or secular travels that shredded principles of fixed property ownership and that mocked a social order in which, as with garden crops, everyone had his time and place. Free-burning fire was an index of social unrest, feeding on the scraps dropped from the high table of civilized agriculture and on the weeds growing through the cracks of crumbling or ill-tended civilizations. Because these conditions were often experienced during rapid colonization, Europe found itself often at odds with settlers in the hinterlands.

Europe's fire philosophes still do not accept such models for nature's economy. The land should serve society: nature's economy should not only support but emulate the state's. Creative destruction by free-burning fire, like untrammeled capitalism, leads to waste, danger, and wild swings of too much and too little, patterns anathema to visions of social harmony, as Europe historically has understood such visions and as those visions manifest themselves today in the ideal of the welfare state. Thus, the immense panorama of burning that persists today in Africa, Asia, and Latin America, no less than the specter of soaring flames in Neo-European nature reserves, challenges European conceptions of how people and land should join. The competing conceptions of fire's ecology did not cease with the colonial era. Outside Europe proper, strange fire continues to profane the Earth's altar.

The global controversy over climatic warming is at core a quarrel over combustion, and it is hardly surprising that Europe argues for a ruthlessly utilitarian and "social" model of combustion for which open flame can find scant opportunity or justification. Europe would prefer that open flame be replaced by a more domesticated internal combustion, which in turn obeys the rituals prescribed by its reigning priestly castes of Eurocrats. Its model is proper; the others, profane. If it could, Europe would blast such strange fires and their sacrilegious servants into oblivion.

a
land
between

The fires rise and fall with the winds. The northerly monsoonal flow off the Atlantic retreats, and the harmattan, parched as its Saharan source, advances south. The rains stop, the fires begin. At first, they creep and smolder, grasping at early-cured grasses and forbs. As the vegetation dries out in larger patches, the flames swell and burn more fiercely. Before the rains return, almost all that can be burned is burned.

The causes are endless, as inevitable as the dry season, as necessary as the sunshine. A register of fire is a litany of rural life. Pastoralists burn to sweep away encroaching brush, cleanse a site of ticks, inhibit snakes, and shock dormant grasses back to life. Their flocks crowd onto the fresh green fodder, more palatable and more nutritious than the unburned stalks. Farmers burn to clean fields of rice and millet stubble, maize stalks, and sugar cane, or, more broadly, the longer-cycle bush fallow of shrubs and trees. For a while, they clear away competing plants, release nutrients, fumigate fungi, and drive off microfaunal pests and weeds. For a year, perhaps two or three, crops flourish. Local languages encode the best time for burning into their names for the seasons. Twi calls February Ogyefuo, literally meaning "fire farm"; Akwapin calls it Apambere, "time of collecting smoldering stems"; and Ewe names it Dzove, which means simply "burn." Farmers in Bawku will ridicule neighbors as "untidy" and slovenly who have not burned before the first rains. Hunters set fires to smoke rats out of holes and both to drive and to draw game. Their flames strip away cover for snakes and flush out mice and grasscutters, sometimes into gauntlets of guns, sometimes into the collapsing circles of communal hunts; later, after the rains, the duikers and bushpigs and antelope, along with the livestock, return to the succulent green pick that luxuriates on the burn. A hunting field typically burns as a quilt of smaller patches, moving game as one might herds of cattle. The pattern of burning also helps shield inhabited sites from cobras, ticks, mosquitoes, lions, or any predator that prefers tall grass and thick scrub for cover. Others burn to gather honey, using smoke to drive bees before tossing the torch to the grass. They set fires under the *dawadawa* trees to promote fruiting; under kapok and mango to protect them from hotter, late-season fires; and around other flora to stimulate medicinal plants. They daily burn out the holes drilled into uprooted palms for the tapping of palm wine. They fire off the Accra Plains to eliminate the grassy dew into which mosquitoes lay eggs. They burn woodlands to kill trees for later use as fuel wood or burn them again to make charcoal. They fell snags by firing around the base. Villagers burn their trash; farmers, their refuse; school kids, tufts of grasses around their playgrounds. Travelers—hunters, fishermen, migrants, sojourners—abandon campfires that escape into the bush, toss cigarettes that kindle savannas, leave a trail of fires like roadside litter. Residents carry fire-sticks or vats of coals from place to place, while embers blow from cracked pots into a fire-eager bush. Fires escape from *pito* brewing at funerals. Prudent forest guards early-underburn through teak plantations to remove excess fuels. Wardens in Mole National Park early-burn to hold the reserve's fabled wildlife within the bounds of its unfenced domain. To the mix of careless burners add the malicious ones, the arsonists and warriors. Add, too, those who burn for fire festivals. In northern Ghana, this practice consists of tossing torches high into selected trees. So extensively is the landscape fired in Ghana that protective burning—burning under controlled conditions to vaccinate against wild bushfires—only clears off what fuels remain. Villagers burn around their grass-roofed houses; farmers around still-unharvested crops, rice fields, and cocoa plantations; priests around sacred groves; and even foresters around gazetted reserves—all early in the season, each with a mind to erect a black, then green incombustible barrier that will last through the harmattan. All these fire causes are not unique to Ghana, but their cumulative effect—the sheer, indefatigable, unsparing penetration by fire into every biotic nook and cranny, its deep symbiosis with everything human—is that West Africa, between the desert sand and the salt sea, has the highest proportion of annual fire in the world. The scene is visible from space, imaged through satellites by day in infrared and by night in visible light, a vast spangled constellation of fires, like a Milky Way of burning, winding across the Earth's dark matter.

Virtually everything people do in the Ghanaian bush involves fire. Remove

those enabling flames and human life would collapse; few things are as fundamental to Ghana's national identity. Appropriately its flag records this environmental saga. Three bands of color fill the field, with a black star in the center.
The bands rise from bottom to top, one leading to another in mimicry of a year's
climatic cycle—green, the season of rains; yellow, the time of dormancy; red,
the fires that follow; the black star, the ash and char that remain at the core of
Ghanaian life and its ambiguous future.

No one can doubt that fire pervades Ghana. But must it? Is fire in some way
"natural" and therefore necessary? Or is it the product of human contrivance
and therefore, at least in principle, expungable? Is it like the sun, about which
one can do nothing save seek shade, or is it more like malaria, for which one
can invent vaccines or hope for outright eradication someday, or is it rather like
the corruption that frequently infests regional politics, for which extinction is an
ideal, worthy if unattainable? It is all of the above, and so inextinguishable.

Ghanaian fire flourishes because both nature and culture favor it. Bushfire's environmental basis resides in West Africa's ancient seasonal rhythms,
with well-defined wet and dry seasons, two such seasons along the coast, one
longer swell in the north. That pattern follows the Intertropical Convergence
Zone (ICZ) as it writhes across Africa, gathering winds from the north and the
south and mixing them until they boil up into daily thunderstorms. When the
ICZ pushes inland, winds with a fetch across the whole South Atlantic bring
rain. When the ICZ slides toward the coast, those moist winds recede, and arid
winds from the Sahara, the harmattan, rush across Ghana. The land dries, the
burning begins. The flames catch the breeze. Ghana with the wind.

A weave of wet and dry stretches across the countryside as well as across
the seasons. The far southwest is consistently wet, the northeast only seasonally
so. A gradient of moisture marches between them such that rainfall decreases,
trees shrivel, and soils shallow as one moves inland. An evergreen forest grades
into a seasonally wet forest, which thins into the wooded Guinea savanna, which
further opens into the edge of the Sahelian savanna. The more-closed forests
resist fire even in the face of drought unless their canopy cracks open and more
fire-prone species invade. Elsewhere the fuel fabric shows a tightly wrought

woof and warp of fine-stranded combustibles, fluffed with expansive grasses, forbs, and other fallow-thriving species. Agricultural practices loosely follow these trends. There is more herding to the north, more tree-crop cultivation (such as cocoa and palm oil) to the south.

This geographic gradient is elastic. Other, longer rhythms compound the annual cycle of the rains. The ICZ is not a geodetically measured boundary but a churning riptide of air masses that wiggles and swings around the globe. How far inland it moves in any particular year and when and for how long are not mechanically fixed. Droughts are common and sometimes profound. In normal years, fire burns in tight patches, only where the dead grasses are thick and the woods felled and cured; the surrounding vegetation remains too moist to carry fire. But droughts crush that difference out of the land, like presses squeezing sap out of cane. Then fires can burn nearly everywhere; they even can creep (under severe conditions) into closed forests. Ghana's geographic gradients become a climatic accordion, alternately pulled and pushed, scrunched between the sea and the sand. In one critical decade, great droughts struck in 1972–73, 1976–77, and 1982–83. The harmattan blew like a furnace fan. Ghana burned inextinguishably.

But fire exists not only or even primarily because natural conditions favor it. Fire flourishes because Ghanaians have put it there. Burning is as fundamental to Ghanaian society and economy as to its biota. Lightning presently accounts for less than 2 percent of ignitions. Almost certainly that number of starts would be higher under a purely natural regime. The explanation, however, is simple. Humans burn first. By the time lightning arrives with the onset of the rainy season, there is typically little left to combust. Thus, whatever affects Ghana's people affects its fire regimes.

In this way, other cycles bonded to social life compound those of climate. There are demographic rhythms, historically tied to wet and dry periods such that the regional population builds up during wet times and collapses during dry (with the surplus populations historically shipped off as slaves). There are economic rhythms, linked to the introduction of new foodstuffs and to the export of commodities beyond regional networks. Of the contemporary

Ghanaian diet, only bushmeat, snails, millet, and yams are indigenous. The rest—plantains, coconuts, maize, cassava, rice, beef, pork, goat, cocoyams, groundnuts, sweet potatoes, and the others—were invented elsewhere and brought into Ghana through trade or migration. Moreover, as long as external markets have encouraged them, its farmers grew for export as well as for domestic consumption such crops as cotton, copra, kola, palm oil, and especially cocoa. All this subjected Ghana's fire-flushed landscape to the uncertain economic winds of trade and capital. Balance of payments shortfalls proved as significant as moisture deficits, global depressions as vital as drought. These distant markets for capital blew over Ghana with the fire force equivalent to the harmattan.

Because in Ghana, as elsewhere, fire follows fuel, the rhythms of cropping and fallowing became the rhythms of bush combustibles. Some patterns remained cyclic, tied to the rains. But new crops, iron tools, fresh livestock (where the tsetse fly permitted), new markets for cultivated commodities—all imposed longer wave patterns, not unlike the ancient ebb and flow of decadal drought. Land given to one use changed to others. In particular, eager farmers cracked open forests to plant crops, while foresters added to the sprawl by culling hardwoods, often in association with cocoa plantations, occasionally to replace the wild woods with teak or palm oil plantations, but always, it seemed, allowing pyrophytic weeds and scrub to invade the now sun-spangled sites. And to these commodities one should add the minerals that dazzled early-voyaging Europeans, who long called the place the Gold Coast. One after another commodity markets rose, then collapsed. They left behind, however, a different landscape, unable to return to the status quo. These long-wave secular changes were equivalent to the south-encroaching Sahara. Together they squeezed and released the gradients of the cultivated landscape in ways that variously clashed, compounded, and canceled out.

Perhaps the most visible reforms were political, for these changes gave shape to national borders, and so defined for a cosmopolitan audience the national identity and have profoundly influenced the last century of Ghanaian life and land. Abundant slaves and gold had dotted the coast with trading posts (forts, really, so tenuous was Europeans' hold on the interior) from Portugal,

Holland, Denmark, Sweden, and Britain. The strangulation of the export slave trade, beginning in 1807, rocked the economic foundations of the indigenous powers. Instead they began to redirect their enslaved throngs to colonize the surrounding countryside in weird mimicry of the plantations erected with slave labor in the Americas. By the nineteenth century, only Holland and Britain remained, and after the final suppression of the trans-Atlantic slave trade, the Dutch withdrew in 1872. A year later British troops smashed the dominant political force inland, the Ashanti Confederacy, centered in Kumasi and declared the Gold Coast a Crown colony in 1874. Europe's unseemly scramble for Africa quickly fleshed out borders, duties, and rights. Twenty-six years later British troops had to return to complete the military conquest. In 1902, the British Crown formally annexed Ashanti as a protectorate to its coastal colony.

The British practiced cost-saving indirect rule. At the height of its colonial presence, probably no more than four thousand Britons oversaw the country. Moreover, endemic diseases and uncertain markets for goods other than minerals and, later, cocoa stalled development. But a new capital, Accra, arose; roads and even some railroads punched through the bush; fresh export crops were introduced; the standard institutions of colonial governance appeared, from forest reserves to a college, a civil service, and a branch of the Royal Society; and both trade and peoples began to move along the great gradients from the interior to the coast. These fundamental, evolutionary reforms were not readily reversed. Then in 1957 Ghana became the first of Europe's sub-Saharan colonies to achieve independence.

Expectations were high as Prime Minister Kwame Nkrumah became the black star of African liberation. Such circumstances made the subsequent political and economic crash all the more damning. Nkrumah's administration of excess and indifference ended in wild inflation and monetary riot followed by a coup. A downward spiral of political and economic recovery and collapse continued until 1982. By then, the average Ghanaian was far poorer than at independence. Virtually every economic index had plummeted. Twenty-five years of autonomy—an era during which South Korea, beginning with the same per capita income, had quadrupled its income level—had in Ghana destroyed wealth. Yet so long as the various rhythms canceled one another out or came in syncopation, the system could muddle along with an episodic coup or default or exchange-rate scandal. Then came 1982–83. The cycles compounded. Ghana crashed and burned.

A new long wave of drought settled over the Sahel beginning in 1968. By 1972–73, its depredations had alarmed the international community. Conditions abated, then worsened in 1976–77, then waned again before returning with massive effect in 1982–83. The drought spread out of the Sahel like stiff molasses. It shoved against the leisurely swells of Ghana's geographic gradients until the desert seemingly pressed against the sea. The drought worsened beyond the scope of written records. Even Lake Volta dried up. The landscape's traditional mosaic broke apart as its normally wet grout dried and cracked. Fires spread like swarms of locusts. Smoke crowded out even the perpetual dust of the harmattan haze. Bushfires incinerated 35 percent of Ghana's cereal crops; they invaded fire-sensitive cocoa plantations; they swept over bush fallow and pasture like flame through Kleenex. They burned villages. They even probed and seared the closed forests. They complicated a vast refugee tide spilling into Ghana from the Sahel and Nigeria, which was forcibly repatriating a million Ghanaians. And they shattered the provisional political reforms of Flt. Lt. Jerry Rawlings's "second coming."

Rawlings had led a coup in 1979 that deposed and summarily executed the corrupt Gen. I. K. Acheampong and several of his cronies. Incredibly, Rawlings had then retired in favor of president-elect Hilla Limann, but social, economic, and climatic conditions continued to deteriorate. On the last day of December 1981, Rawlings removed Limann and installed himself as head of the Provisional National Defense Council (PNDC). A year later the PNDC announced a four-year economic plan. But the Soviet bloc informed him that it would give nothing, and drought continued to strangle the countryside. Money and rain blew away on the desert winds. With Wagnerian sweep, the fires ended Ghana's political opera in a flaming coda. Rawlings turned to the International Monetary Fund (IMF) for advice, and the country began to rebuild its economy along more transparent and market-friendly lines until, a decade later, it had again become the economic black star of Africa.

The association of fire with Ghana's Götterdämmerung made the reform of Ghanaian fire seem a natural part of reforming Ghanaian society overall. As the new government sought to integrate Ghana into the global economy through the advice of the IMF, so it also sought technical help for coping with bushfires. The U.S. Forest Service sent advisors in 1985. In 1998, the International Tropical Timber Organization launched a three-year program to stabilize fire threats along the cordon of fraying forest reserves that separate the high forest from the savanna. Other international donors, including the Netherlands, proposed projects to improve bushfire protection. But exchange-rate mechanisms proved easier to reform than fire practices.

Ghana is far from unique in the fire saturation of its rural landscape. Begin with the observation that until very recently almost all agriculture depended on fire for both its creation and its sustenance. Yet there are parts of the world for which open burning has become expendable. Why? The simplest answer is that they industrialized. They no longer have an economy that depends on agriculture, or they possess an agriculture that no longer depends on fire. They broke out of—transcended, really—the endless cycle of growing and fallowing by tapping fossil biomass, which serves as a kind of fossil fallow. They still reach beyond the growing field for fuels; they still combust them. But they burn or distill coal and oil, and they do so by indirect means—by tractor-drawn plows and harrows, by artificial fertilizers, by pesticides and herbicides—so they no longer fallow and no longer openly burn that rank growth. In such places, rural fire has faded away.

That is the starkest explanation of how countries have moved beyond a rural fire regime. The problem of promiscuous rural burning was never really solved in any technical sense except in those places for which fire existed only through willful application. So long as a rural economy flourished, so long as a rural population committed to fire-fallow agriculture resided on the land, so did fire. Something had to move the farmers and herders off the landscape. Industrialization did that in part by reducing the number of agricultural laborers needed and in part by creating jobs in cities. Often, however, the rural population was removed forcibly through enclosure and collectivization or in colonial situations by disease, war, and eviction. All were violent processes; socially violent, ecologically violent, or both. None offer suitable models for today's developing countries.

The point is that rural fire control occurred at a local level or not at all. At the level of a national or even provincial government, there has been no successful "high-modernist" strategy for coping with rural burning. The solution to excessive burning historically was to regulate it better locally and to evolve nationally beyond an economy based on fire-fallow agriculture. But however that metamorphosis occurs, it tends to be painful, and the era of the pyric transition overflows with fires. The United States, for example, experienced horrific holocausts during railroad-powered settlement in the latter quarter of the nineteenth century. This abnormal scene typically alarms officials, outrages intellectuals, and hardens institutions in a commitment to abolish fire altogether. Even so, U.S. officials were unable to solve the problem in any systematic way. The prevailing belief was that the fires would vanish as settlement matured, that forest fires would disappear as forests became farms. The cycle of abusive burning continued, in fact, for more than sixty years and ceased only after the rural landscape was exhausted and more or less abandoned. (The ordeal has loosened its grip much more slowly in the South, where a rural population persisted and fire flourished despite the edicts of politicians and the exhortations of conservationists.) Instead, the successful "firefight" occurred on reserved public lands in the United States—precisely those landscapes not subject to the pressures of a rural economy.

There is no easy solution to Ghana's rural fire economy. If history and analogy are useful as guides, Ghana will not control bushfire through legislation or force or persuasion. The current fire regimes will persist until Ghana evolves out of them. The critical issue is how to pass through that fiery metamorphosis without savaging the landscape beyond redemption. In particular, it is an open but vital question whether Ghana's forests can survive a prolonged pyric transition.

The burden of answering that query has fallen to the Forest Department. This being Ghana, whatever might threaten the forest will somehow involve fire and will likely do so on such a scale that the control of fire will appear as an end in itself instead of a means. An exercise in fire exclusion would seem quixotic except that, incredibly, fire-free forests do exist. Unsurprisingly they have become objects of intense scrutiny for what they say about the politics and ecology of fire protection and thus for what they might promise for the future of the Ghanaian forest.

The best known are sacred groves. They are not many, and they are not large, either individually or collectively. But they demonstrate that it is possible to shield woods from fire if the local community wishes to do so. Some human access is allowed with priestly oversight, and some harvesting of special plants may occur (and may provide one of the unpublicized reasons for protection). The rationale behind fire exclusion is special, however. The primary purpose is religious or, more broadly, the identification of a particular site with a people's origins; usually these sites commemorate the burial of a group's founder or its ancestors or in some other way lays a claim to social memory. For this reason, the preserved woods are themselves a relic, the biotic equivalent to the symbol-charged "stools" that distill the tribal story and define political legitimacy.

A second fire-free zone occurs around the Nandom District in far north-western Ghana, specifically the villages around Gozrii. Here the local chief, after suitable consultation with his council of elders, banned burning. One site of several hectares has now escaped fire for thirteen years. The understory has roughened, the woods have thickened. Fire critics have advertized the Gozrii experiment as a model for what might be achieved elsewhere or even across all of Ghana. Yet its success seems to depend on special ecological and political circumstances that are unlikely to generalize beyond the personalities of the paramountcy. The moral is simply that it is possible to exclude fire, at least for a number of years, provided the local community has the will to do so.

That leaves Ghana's secular groves and its reserved forests under the direction of the Ghana Forest Department. The reasons for the removal of fire are again both environmental and political. The reserves dapple a region that promotes high, often closed forests for which fire is not a natural presence and that, in fact, repel bush fires that strike along their dark border. The woods thus enjoy a certain intrinsic immunity from burning. That they exist outside the matrix of Ghanaian agriculture also excludes many ignitions. The political explanation is that they are overseen by an agency of the national government committed to fire protection. For most of the twentieth century, these two processes have collaborated to shield the reserves from routine fire. In all, the reserves legally protect approximately 20 percent of what were once Ghana's high forests and are all that remain outside of parks.

More recently, both the ecological and the institutional shields have frayed badly. The Forest Department is evolving into the Ghana Forest Service, which is committed both to a strategy of collaborative forest management and to a program of self-financing its operations. The outcome is ambiguous; at a minimum, the authority as well as the bureaucratic presence behind state fire protection have receded. It is not obvious that these institutions will survive or, if they do, that they will exist as more than paper bureaucracies. Still, the reserves probably could endure these political strains were they not also stressed simultaneously by environmental factors. Forest reserves now exist as islands in a sea of agricultural fire, and climatic storms (like those of 1982–83) are rapidly eroding the exposed biotic coastlines. Most of the reserves have suffered from logging, agricultural encroachment, weed infestations, and fire; some have degraded significantly. The shock is particularly acute along the so-called transitional zone between the savanna and the high forest. Whatever their legal status, the reserves nonetheless hold most of what endures of Ghana's high forest in a form that approximates a coherent biotic system.

It is small comfort to thoughtful Ghanaians to know that their problems are again not unique. The creation of forest reserves was a colonial practice of the nineteenth century and continued for a few decades into the twentieth century. Ghana's reserves were among the last established by Britain, but for that reason they represented a degree of acquired wisdom and, for the Gold Coast, a compromise with local politics. Several false dawns appeared between 1907 and 1912, rudely based on the Indian model and assisted by an indefatigable forest officer, H. N. Thompson, loaned by Nigeria. The process collapsed under the pressures of the Great War and local protests, and the office of Conservator of Forests was abolished in 1917. An acceptable new policy emerged in 1919; officials gazetted reserves; and in 1929 a revised forest ordinance accelerated the movement, which more or less achieved equilibrium in 1939. Yet ultimately the reserves remained an imperial institution: they expressed outside ideals and existed because of the power of the colonial authorities to install them.

Their origins, however, do not invalidate the reasons for their creation. In part, they existed to rationalize the timber trade. Experience repeatedly showed that local institutions collapsed when suddenly confronted with a heavily capitalized global market. Concessionaires typically scalped timber and left. Foresters promised to regulate those practices and to repair the damage done; the reserves were models to demonstrate how to accomplish this task. They were a means by which the state could interpose itself between local and global processes, to protect the former and promote the latter.

Equally, reserves existed to bolster larger, national values, as secular equivalents to sacred groves. It was believed that protected woods could ameliorate climate, help regulate navigable rivers, even provide a degree of solace for harassed urbanites. They spoke to national, even imperial values—the needs of a larger world—that local communities could not understand and would not support. Control thus had to reside outside the hands of tradition-bound peasants, transient hunters, and transhumant herders. In brief, reserves prevented economic monopoly, retarded the ruin of natural resources, and promoted transcendent goals. And, it is important to note, they often worked more or less as planned. Without its legacy of reserves, Ghana today probably would have no standing high forests of any consequence or coherence.

Nature reservations, most expansively forest reserves, appeared throughout Greater Europe's imperium. They were most successful where European settlers displaced the indigenes, less so where Europe ruled over local peoples. In the former case, the reserves were often uninhabited. They were created at a time when the indigenes were gone but European settlers had not arrived in force. The lands were, in a profound sense, vacant, and they remained the property of the Crown, province, or state. Where peoples persisted in or around the reserves, administration was complicated and compromised. In a few instances, colonial foresters administered the reserves on behalf of local rulers—as was the case with Ghana and with American Indian tribes. The outcome is probably the least politically stable of all such arrangements.

It is not obvious how to protect such lands. They cannot be reserved without being used, cannot be used without introducing fire, cannot be burned without changing. Although bushfire is neither the source nor a driver of such change, it is a necessary catalyst, and unless it can be controlled, the reserves will erode away into weed lots and bush fallow. Without fire protection, legal protection is meaningless. Ghana's relationship to its forest reserves, as to the rest of its land, is one mediated by fire.

Bushfires have entered the reserves proper because logging has cracked open their multilayered forest canopy and then allowed pyrophytic weeds to invade. For those reserves dedicated to production forestry, logging has started a conversion that has replaced native woods with commercially valuable exotics. But whether the old forest is allowed to recover or a new one replaces it, fires are now an inextricable part of the system. They have entered most reserves and established residence. So long as logging continues, so will fires. Logging prescriptions must now factor in fire protection, and timber harvesting must accept those expenses as part of its cost of production even as the relative value of timber among Ghana's export commodities slides downward. A complicating factor is that the new Ghana Forest Service is expected to pay for itself through commercial logging, and as the local stools receive a greater fraction of the accrued revenue, both the incentives for logging and the dangers from bushfire will ratchet upward.

Outside the reserves, the problem is one of fire-fallow agriculture. So long as rural life dominates Ghana's economy, fire will scour and polish a domesticated bush. Inevitably some of those flames, however, will crash into the reserve border and splash across as wildfire. Yet neighboring villagers are also the only means for bushfire control. They alone can prevent fires, fight fires, and rehabilitate after fires. They offer the only reliable source of staffing for bushfire fighting, though they have had little incentive to support the Forest Service and many reasons to resent what they consider the appropriation of "their" lands by an alien bureaucracy. The redirection of forest policy into collaborative management—obviously modeled on community-based natural-resource management programs developed for wildlife elsewhere in Africa—offers a partial solution. It is conceivable that villagers will do for the Forest Service what it cannot do for itself, provided sufficient incentives are offered. Further compromising civil authority are the swarms of nongovernmental organizations that seem to broker every new enterprise, the political equivalent of a black market. Political authority promises to become as complex and intricate as Ghana's polycropped farmlands.

The strains are greatest along the borders. The reserves are (to put it kindly) weirdly constructed or gerrymandered outright. Most reserves have a very long perimeter relative to their bulk, which makes them even tougher to defend. At least along the transitional zone, most borders are not well demarcated. They exist as an occasional sign and a cutlass-slashed swath through the annual flush of grass and scrub. They offer nothing to halt trespass or fire. They rely, ultimately, on the goodwill and discipline of the surrounding villages. The decay of their boundaries, such as those at Afram and Pamu-Berekum, into fields of elephant grass and *acheampong* dissolves even that presumption. (Acheampong is a tenacious, exotic pyrophyte named after the deposed dictator.) Without a border, the reserves have no identity. Yet the Forest Service lacks the resources to defend those boundaries, and villagers the incentive to assist them in the task.

The proposed solution—ingenious, really—is to construct a system of "green firebreaks" around the threatened perimeters: "green" because they consist of incombustible crops; "firebreaks" because, at forty meters in width, they should halt a spreading surface fire; "ingenious" because the Forest Department can enlist villagers to do the work in return for new cropland. Because the planting includes fast-growing, sun-blocking species, it is assumed that after three to four years the trees will overtop the abandoned fields, and the fuelbreak will become sufficiently shaded and self-sustaining both to halt fires and to waive the need for future maintenance. In principle, a program of such fuelbreaks might staunch the further fire hemorrhaging of the reserves. It would leave the reformed Ghana Forest Service with administrative control and the villagers with practical access. The Netherlands has proposed a ten-year project around just such a conception.

Fuelbreaks are rarely more than a partial solution, however. They tend to serve only as an enabling device. They work best when they are built into the design of a site, not retrofitted. They help move a land from one status to another. Their assets are that they assist in the control of surface fires and that they are supremely legible to observers, whether critics or advocates. They are an obvious expression of an administrative presence. There are instances where they have succeeded—in taming prairie fires, in replanting burned forests, in protecting pine plantations—yet they often look better than they perform. Their liabilities are formidable. They are expensive to install, especially when imposed on a landscape originally fashioned for other ends. They are even more expensive to maintain. Because they consist of plants, they grow, change, evolve, and require tending; most fall into disrepair when their informing crisis has passed. They fail during extreme fire behavior, which is precisely when the need for protection is greatest. Spot fires soar over them with complete disregard for the labor and cleverness invested. And, perhaps most tellingly, they tend not to be permanent features. They are transitional, and they suffer when the larger goals are inappropriate or unreachable.

The core problem remains political. If fuelbreaks work best as transitional devices, then what is the end result, and how long will it take? The critical transitions are several. There is, first, the forest itself. Assuming the scheme does exclude fire, at what point has it succeeded? Particularly if logging continues, as it must to finance the Ghana Forest Service, then the fuelbreaks must remain more or less permanently, which will create serious difficulties with maintenance.

The scheme is, second, transitional within the evolution of the Ghana Forest Service. The institutional arrangements will be redefining themselves for many years to come, perhaps over decades. The role of the Forest Service as a viable agency will depend on the definition and the perceived value of the reserves it oversees. Without those reserves, there is little justification for a forest bureaucracy within a state bloated by civil service patronage. But as the reserves are used progressively, so their purposes have multiplied.

The trade-offs are tricky. The reserves' diversity of purposes dissipates as well as concentrates their political rationale. If the reserves serve only one purpose, they will lose constituencies. If they serve too many, they risk a loss of definition, and even as the agency attracts some clients, it alienates others because, at particular moments, it must choose one or another. The Ghana Forest Service simply may absorb the conflicts within itself in perhaps bureaucratically painful ways. One solution is to subsume the reserve system under an umbrella organization but subdivide the various functions among separate bureaus. The Forest Service, for example, might have special interest in commercial logging and plantations—the direction toward which the institutional momentum seems to be headed. It may be that such definitions, like the proposed fuelbreaks, are a provisional device.

The most powerful transition is, of course, that of Ghana from a developing, rural nation to an industrial one. This is not something the Forest Service can create, though it will be affected profoundly by the outcome. When the rural economy subsides, so will fire pressures on the reserves. Any scheme of protection must see the forests through that phase, which, barring major setbacks, will likely last for forty to fifty years. A scheme also should provide protection under the worst fire conditions, such as the return of a season like that of 1982–83. One or two equivalent outbreaks (a likely outcome over the first half of the twenty-first century) and the highly exposed reserves simply may burn away, save those in the most remote and damp locales. Climate and culture, both unpredictable and both outside the control of the Ghana Forest Service—how they mix or mash will determine the timing of large fires, and the outbreaks of those fires will determine the success of forest protection.

Paradoxically perhaps, the multiplication of purposes behind the reserves by itself may not add up to a compelling argument. If the forests are valued primarily for timber, then the state may seek to privatize them and international donor nations may lose interest in subsidizing their administration. The values that may be gained are too ephemeral and those values threatened with loss too remote. Environmentalism behaves as though on a green mission to civilize Africa that has replaced for many affluent nations the antislavery campaigns of old. Remove that incentive, and the public appeal will fail. If the reserves exist to promote biodiversity or ecotourism, then a parks or wildlife agency may be the most suitable instrument. If the reserves belong to the surrounding stools, and those peoples want to determine for themselves what should be done with land, then the Forest Service becomes a bureaucratic barrier, one that is politically inexpedient. The reasons for allowing the reserves, which are after all a colonial relic, to fall away are many.

Yet a compelling logic exists for holding them more or less intact. It resides not in any one purpose or use but in the simple fact that the reserves help keep Ghana's environmental options open. If the reserves collapse, then so will Ghana's relic high forests. They will disappear into stump fields and farms, and once gone they will not be restored. They were the creation of a unique period of Ghanaian history. When Ghana emerges through its industrial adolescence, it will wish it had those reserves, for purposes it cannot today define, and the only way to ensure that those woods survive as biological landscapes, not simply as hollow legal entities, is to protect them from fire. The alternative is for the country to find that the winds unfurling its symbol-laden flag have scattered its woods, that its forests have gone with the flames riding the surge of the harmattan.

How bushfire protection will happen is difficult to predict. The uncertainties are too great; the complex tables of programmatic elements, detailed strategic plans, and elegant policies and declarations that appeal to international donors and IMF advisors are almost certain to be frustrated. Most likely the Ghana Forest Service and its sister agencies will muddle through, though in a vigorous and creative way. They will improvise. They will find ways to tend the necessary fires and contain the damaging ones. They will cross this period as one might hop from log to floating log across a pond, not stopping long enough to tumble over. With pluck and luck, they may carry most of the reserves along with them. A century from now Ghana will be glad they succeeded.

ecological
imperialism,
and
its
retreat

Half the species in New Zealand have originated from somewhere else. The
two islands are chockablock with exotics—flowers, grasses, trees, shrubs, birds,
insects, mammals, fish, microbes. World-record trout flourish in Lake Taupo.
Red deer thrive on ranches. Himalayan *thar* scamper among the peaks. Gorse
and blackberry line fields and overrun hills. Monterey pine sprout in thick rows.
Its entire agricultural output of course derives from wheat, corn, white clover,
sheep, and cows introduced by eager farmers and pastoralists from across
the far seas. Yet much of the pre-European and a chunk of the pre-Maori
biota endure, from majestic kauri to secretive kiwi, if sullenly and pressed to
remote niches. The place is a historical menagerie of ecological history—or,
as it has become for environmental historians, a model study of biological
colonization. In fact, thanks especially to Alfred Crosby's zesty, wry scholarship,
it has emerged as the very exemplar of Europe's ecological imperialism.

Britons commenced the serious colonization of New Zealand in the mid–
nineteenth century, which places the experience squarely within the flood tide
of European expansion. But the past fifty years have witnessed the ebb of
empire. Even as environmentalists rail and historians unearth more evidence of
Europe's biotic penetration into an "indigenous" world, the facts on the ground
testify to imperial recession. New Zealand has redefined the Treaty of Waitangi
into the basis for a formal constitution. It has begun transferring lands back to
Maori jurisdiction, if not outright sovereignty. It has striven to define itself as
a Pacific nation, with stronger links to Polynesia than to Britain. It campaigned
vigorously against the further introduction of exotics. In 1984, the Labour

government of David Lange started dismantling much of the moribund institutional apparatus of the state, a good bit of it inherited from its colonial master, Britain, and indeed did so at roughly the same time as Thatcherite Britain did for many of the same reasons. The Commonwealth became less a postimperial connection than an institutional shadow, particularly as Britain's absorption into the European Common Market severed preferences for New Zealand farm stuffs and dairy products and as Britain's halting absorption into the European Union realigned its political energies away from the Pacific. New Zealand reconstituted itself from Britain's off-shore market garden into Asia's off-shore tree farm. And it disestablished the New Zealand Forest Service, selling off the choicest commercial morsels and tossing the high mountain catchments into an underfunded Department of Conservation. The story has become one of colonial and ecological recession.

All this has repercussions for New Zealand's fire history. Anthropogenic fire is as much an exotic as gorse and as indigenous as bracken. The country's realignment chisels into bold relief the major drivers of the modern world's fire geography. And it underscores, in particular, the role of imperial institutions in shaping the last century of fire policy and practice.

Ancient New Zealand had never flourished as a fire-driven landscape. Its temperate climate was not prone to powerful wetting and drying on a routine basis. Great El Niño–Southern Oscillation events catch the islands in their climatic sloshing, and the islands' eastern flanks in particularly display foehnlike winds as fronts pass over the soggy Southern Alps and, drained of moisture, rush over high mesas and down valleys below. Charcoal buried in soils and lakebeds records several major outbreaks of burning in prehistoric times (for example, between 2,500 and 1,500 years before the present). Volcanically riven North Island literally boiled with disturbances, including no doubt some burning. But regular fire and fire-shaped regimes had to await human contact.

This contact came with the Maoris, perhaps A.D. 1000–1200. They transplanted a fire-fallow agriculture on North Island and applied fire hunting wherever they could, with notable success on South Island. Many forests evolved into long fallow or converted to bracken and tussock. The moas disappeared. The habitat morphed and became, thanks to human finagling, more fire prone. The Europeans quickened the trend. They brought a veritable ark of flora and fauna. Many of the plants were pyrophytes, particularly weeds. Most of the animals were grazers and browsers, which helped trample and chew the biota into more fire-malleable forms. During periods of exceptional land clearing and drought, great fires broke out. The 1885–86 fires around Hawke's Bay (and elsewhere) were part and parcel of a global swarm of deadly burns that accompanied the violent transformation of the land to better meet European norms. But everywhere—large and small, fallowed field or bracken patch or land scalped by logging—fire acted as a catalyst.

Then the fires abated. As the landscape assumed more or less stable form, primarily as close-tended farms and wide-ranging paddocks, the fires, too, settled down. They became a feature of rural life, common and useful if not quite wholly respectable. In 1921—after several abortive attempts, but still rather late by imperial norms—New Zealand created a Forest Service. It promptly began wrestling the unfarmed bush into better order. Fire protection loomed large in its vision. And as happened generally, forestry became a mechanism for installing fire services over a wider rural landscape. In all this activity, New Zealand followed a broadly imperial model, which is why the disestablishment of its Forest Service seems so shocking and why that dissolution has rippled through the landscape of rural fire.

Modern New Zealand can stand for most "settler nations" in that its fire history braids together three narratives. One, the deepest, is an industrial narrative. The wholesale substitution of fossil for living biomass has transformed not only the nation's economy but its intricate landscape. Urban and rural New Zealanders need flame less, so use it less. The extinction of open burning is simply an index of the country's larger economic trajectory. The industrial revival of New Zealand during the 1980s is both cause and measure of fire's removal. The night scene shows more lights and fewer flames.

The second strand is an imperial narrative. The demographic takeover by British immigrants created often large expanses of depopulated lands. These lands the colonial governments claimed for the Crown. The juiciest swaths included forests—their reservation was, in fact, approved in the name of the common weal, as a bulwark to stabilize climate, rivers, soils, and the like. Throughout the British imperium, such lands fell to the oversight of foresters. Forest services were thus, in principle, far more than timber companies, although they did plenty of logging. They were an imperial institution justified for the service of a larger good. Everywhere they became a prime vehicle for suppressing fire.

Yet each country had its own peculiar politics and land use, which leads to the third strand, a national narrative. New Zealand was unusual amid the Commonwealth in that it was not a nation confederated out of several colonies but was managed as a single entity, which meant its forest reserves did not ultimately devolve to the provinces, as they did elsewhere. The national Forest Service prevailed. Indeed, it thrived as an economic organ owing to massive afforestation programs and had become a prominent vehicle for government work projects to sponge up the otherwise unemployed. Throughout, the Forest Service not only fought fires and helped organize rural bushfire brigades but, like the nation's farmers, exploited fire rather as a means of cleaning up logging debris and readying sites for planting.

What makes the New Zealand story intriguing overall is the way that the first narrative, the need to reindustrialize, and the third narrative, the oddities of the local landscape, combined to abolish the second. Alone among the Commonwealth countries, New Zealand formally broke up its Forest Service and thus scrapped all the various fire-service duties it had performed for the countryside. It quickly became apparent how deeply indebted rural New Zealand was to state-sponsored forestry for an infrastructure to protect against bushfires and how many burning practices had depended on it. When the Forest Service dissolved, so did the grout of rural fire protection, and so did the incentives to exploit controlled burning as a means of forest management. When industrial forestry took over, it scotched burning as too dangerous, too imprecise, too likely to arouse public ire (and litigation) over smoke and escapes. The commercial incentives are to apply herbicides to control brush, to prepare sites with machinery and fertilizers, and to rely on rapid detection and attack through heavy machinery to control outbreaks of bushfire. Without the moral force of government to oversee values other than those nurtured by the market and without its partial immunity to tort claims, the justification to apply controlled fire vanished. The state no longer shielded such lands from the full brunt of industrialization and its fire abolitionism. On both counts, the New Zealand saga merits deeper consideration.

The Earth's firepowers today are those that hold extensive wildlands under public ownership. Americans, especially, so take this fact for granted that they rarely question how unusual and tenuous these circumstances are. They argue that "fire is natural" without appreciating that natural fire displays its power very largely because it occurs on public lands, which exist from a convergence of historical accident and political choice. So, too, almost all the debates over suitable fire policies—over the proper proportion of fire control to fire use, for example—assume a sweeping landscape in which to fight or light fires.

The New Zealand experience should remind us that the public domain and the institutions created to oversee it are neither inevitable nor immutable. Their existence is negotiable. They might very well either disintegrate or morph into quite different shapes. Yet the future of fire—certainly its areal dimensions and symbolic power—will depend on the future of such lands and agencies. They are not part of an environmental constitution, a Basic Law, that will stand in perpetuity. The forces arrayed against them are real.

The simplest forces to identify are the pressures to privatize—if not to sell off the land outright, then to outsource the various tasks needed to administer them. There is no reason to exclude fire management from this trend. An alternative is difficult to imagine only because of the U.S. experience that fire protection must be a uniquely governmental duty. In fact, partial privatization already exists: air tankers, helicopters, and catering services currently labor under contract, and a proliferation of consultants has dimmed the unquestioned assumption that only the government can cope with fire. Strikingly, privatization is encroaching on research as well. New Zealand privatized outright its Forest Research Institute, as has South Africa. The other firepowers either have shrunk operations or have begun to outsource more and more of their science.

Equally, there is a pressure to devolve. In much of the Commonwealth, this devolution commenced in 1930, as imperial or national forests were transferred to the responsibility of the provinces or states. The land remained public but came under another level of jurisdiction. This, too, remains an ongoing theme in American politics. Those states that hold the greatest fraction of federal lands are those that were admitted between 1870 and 1912. Earlier, the lands were largely sold off before state-sponsored conservation became gospel; later, states such as Alaska managed to retain significant fractions as state, not federal, lands. (Interestingly, the loser in the Alaskan land scramble was the U.S. Forest Service. Commercially useful lands went to Alaska or native corporations; lands most valuable for wildlife or wilderness went to the National Park Service and the Fish and Wildlife Service—which is fundamentally what has happened in New Zealand.) There is no intrinsic reason to believe that these proportions will remain immutable. Even if federal ownership is retained, the effective administration of such lands will likely become a condominium jointly exercised with local communities. Certainly this is a deliberate policy with regard to the intermix fire. The era of imperial rule is ending and with it an era of fire programs conducted on that model.

Not least, there are pressures literally to recede lands to the indigenous peoples who were on them before European settlement. This is happening not only in New Zealand but also in South Africa and Australia and is in full flush in Canada with the state of Nunavut as well as with the many treaties being negotiated, most spectacularly in British Columbia, that will decide levels of sovereignty over what were previously Crown lands. Again, the imperial state is ebbing, and taking control over fire practices with it. The two primary exceptions are the United States and Russia—Russia because of its lingering reluctance to privatize land; the United States because, as a nation not an empire, it worked more ruthlessly to extinguish native claims. This process, however, the United States had the chance to review in Alaska and there chose to cede vast lands to native corporations. There is little reason to believe that other public lands are not likewise vulnerable to legal and political challenges.

In brief, forest reserves lie amid the deep berm left by Europe's rapid ebb of empire. They are, in the sweep of history, recent creations. The oldest reserves (those in India) have endured for less than 150 years; the youngest, like those in Ghana, for about half as long. Many reserves in former colonies exist only on paper. Most are undergoing some degree of reconstitution. They have survived best where the national government remained strong, the reserves uninhabited, and their perception as a colonial hangover weak. New Zealand's lasted a fleeting 75 years. Yet these are the places where wildland fire most thrives and where it is most astutely studied and where it commands the attention of the state.

The New Zealand experience should remind us how much our conceptions of "fire," of its "place in nature," and of suitable "policies for fire control and use" depend on an exceptional history. Most of the burning in the United States today is where it has always been, in the Southeast. Yet the perceived "fire problems" are those in the West, where the bulk of the public lands reside. We are so accustomed to these circumstances and so assured that their conditions will endure that we no longer recognize them as peculiar or appreciate their many vulnerabilities. Yet decolonialization is as prominent a theme in fire history as is industrialization.

Of course, the American West burns because environmental conditions make it marvelously fire prone. Fires will no more vanish from the Mogollon Rim should the land be privatized than will floods along the Red River should its floodplains become public and the inundated towns be relocated to higher ground. What will disappear is the sense of problem and of public responsibility, if not stewardship, for their solution. However bumbling the policies that have guided U.S. fire history, they have left fires on the landscape. Government ownership has shielded the lands from the full brunt of settlement and industrial conversion and has thus removed the deep coercions to abolish open flame. The prevalence of fire—even if wild rather than controlled—is as much attributable to the dynamics of land ownership as to climate. With or without lightning, droughty or rainy, the American West lands will exhibit a dramatically different fire scene should those lands pass from public ownership or control through the relict institutions that formerly oversaw them with imperial verve.

This seems to be happening. Over most of the public domain, there is less fire than before the lands were reserved, though more than should the land pass into private hands. The fact that the lands were reserved mattered more than that they were subject to aggressive fire control. Despite the urgency of calls to "restore" fire, the fact is that it *has* endured. It could not be abolished by state-sponsored protection and was exempt from the commercial and industrial forces that, where unleashed, have substituted for or suppressed outright open flame. What the mandarins of British forestry termed "fire conservancy" ended not with the abolition of fire but with its preservation. Fire simply proved too difficult to extinguish in fire-prone bush, and, with deep irony, the attempt to exclude fire proved self-destructive. In the end, reserving large forests only created a flourishing habitat for fire.

Critics rightly can point out that there is less flame than there ought to be for lands set aside as natural preserves, and the pressures will mount to remedy that imbalance. Almost certainly the upshot, however, will still be far less burning than existed a century or two earlier. And, with an irony so compounded it turns back on itself, the practice of open burning likely will succeed or fail according to the extent that the much maligned and fumbling state can shield those lands from the pressures that have transfigured fire everywhere else. The privatized forests of New Zealand quickly shut fire down. The public forests of the United States must continue to grapple with free-burning flame.

The model for the future lies less with a national or imperial Forest Service than with institutions such as the Nature Conservancy that can make decisions without the gridlock of national politics and that act on the ground, nimbly and precisely. That model argues for a tinier realm of fire than has existed in the past and one without the romantic sweep of global empires, even empires such as forestry. It also suggests that fire can find less-cumbersome niches. The future seems to belong with smaller programs, with site-specific projects, with solutions devolved to local communities, with work done by private contractors, with compromises between a biota's craving for fire and industrialization's rage to contain all combustion, with the recession of the apparatus of imperial conservation.

The whole saga of "fire conservancy" is thus deeply paradoxical. The attempted extirpation of fire is what, in the end, preserved it. So, too, the effort to restore fire may succeed but at the cost of fundamentally altering the public lands that serve as its habitat and the public agencies that previously have decreed policies. Private reserves, nongovernmental organizations, and consortia such as the Malpais Borderlands Group may prove more effective than national, quasi-colonial organs of the state. Overall, the imperial retreat promises to prompt as much ecological change as the preceding imperial advance—which should surprise no one. The history of ecological imperialism has been written with an acidly ironic voice. There are ample reasons to include fire within that vision. We know the imperial story: what its removal might look like is less clear. If we hold to the ironic narrative, then we can imagine fire's attempted restoration, but amid conditions that destroy its long-term perpetuation. That is, by removing the imperial buffers, the lands might lose their suitability for fire just as administrators seek to reinstate it. The ironic narrative would read: the imperial advance sought to suppress fire and ended up whipping it into more ferocious forms, whereas the postimperial retreat, seeking to reinstate fire, may conclude by encouraging its extinction through competition with industrial combustion. Irony becomes an endless Möbius strip.

Better would be a postironic vision, in which we do not simply invert the old story but invent a new one: fire returns, the public lands change character, irony becomes irrelevant because the imperial past is not the standard. The postimperial era becomes not merely one of wholesale retreat and restoration (which leaves us locked in irony) but simply one of metamorphosis. In this version, free-burning fire returns not by thrusting torches into the existing lands as though the imperial era had never existed or in the hopes that kindled fires will purge the toxins of the colonial era's former presence, but by reconstituting those lands. The resulting fire regimes may little resemble precontact regimes, though they almost certainly will be more robust than those that subsist today. Instead of restoration, we have regeneration. Instead of irony, we have paradox.

That future already may have begun. Small islands feel tremors and biotic quakes more vigorously than continents. We could do worse than watch the New Zealand scene. The saga of ecological imperialism's retreat has only begun. Whether its way is lit by a torch or a flashlight largely will determine the future of free-burning fire.

extended
attack

the

source

It may be the most obscure site on the National Register of Historic Places. Rockfall and wild growth clog the entry. The West Fork of Placer Creek splashes a few feet below. It is not an easy place to find. It has the feel of some mythical grotto, a sepulcher, an oracle, the source of a sacred spring like Lourdes. The Nicholson adit is, in truth, all these, for here, on August 20, 1910, flames burned through conifer stands like prairie grass and came over the ridges, as one survivor recalled, with the sound of a thousand trains rushing over a thousand steel trestles. One ranger said simply, "The mountains roared."

The trek to the site is arduous, not because it is long (it isn't), but because the primary trail, which used to trend to Striped Peak, is abandoned and overgrown, vanishing into a Northern Rockies hillside beneath rockslides, talus, roots, forbs, the slender shafts of willow and alder. A secondary path to the old mine is even more obscure. You need a reason to come here, and you need a tool. You need something sharp to slash through the scrub. You need something durable to grub out steps through the loose rubble and root-clogged slopes. You need a pulaski.

What happened that astonishing summer was that American society and American nature collided with almost tectonic force. Spark, fuel, and wind merged violently and overran 2.6 million acres of dense and odd-disturbed forest from the Selways to the Canadian border. The sparks came from locomotives, settlers, hobo "floaters," backfiring crews, and lightning. The fuel lay in heaps alongside the newly hewn Milwaukee railway over the Bitterroots, down the St. Joe Valley, across hillsides ripped by mines and logging, and through untouched woods primed by drought. The Rockies had experienced a wet winter but a dry spring that ratcheted, day by day, into a droughty summer, the worst in memory. Duff and canopies that normally wouldn't burn now could. The winds came with the passage of shallow cold fronts, acting like an enormous bellows that turned valleys into furnaces and sidecanyons into chimneys. Southwesterly winds rose throughout the day to gale force by early evening and then shifted to the northwest. Perhaps 75 percent of the total burn occurred during a single thirty-six-hour period, what became known as the Big Blowup.

That summer witnessed the first great firefight by the U.S. Forest Service. As the weeks wore on, the fires crept and swept, thickening during calms into smoke as dense as pea fog, then flaring into wild rushes through the crowns. The fledgling Forest Service, barely five years old, tried to match them. It rounded up whatever men it could beg, borrow, or buy and shipped them into the backcountry. The regular army contributed another thirty-three companies. The crews established camps, cut firelines along ridgetops, and set backfires. Over and again, one refrain after another, the saga continued of fires being contained, of fires escaping, of new fronts being laid down. The Big Blowup shredded it all. Smoke billowed up in columns dense as volcanic blasts; the fire's convection sucked in air from all sides, snapping off mature larch and white pine like matchsticks, spawning firewhirls like miniature tornadoes, flinging sparks like a sandstorm. Crews dropped their axes and mattocks and fled. That day seventy-eight firefighters died. One crew on the Cabinet National Forest lost four men; one on the Pend Oreille lost two; the rest of the dead fell on the Coeur d'Alene.

The Coeur d'Alene was ground zero. In the St. Joe Mountains between Wallace and Avery, some eighteen hundred firefighters and two companies of the Twenty-fifth Infantry manned the lines when the Blowup struck. A crew north of Avery survived when Ranger William Rock led them to a previously burned area, except for one man who, panicking, shot himself twice rather than face the flames. A crew on Stevens Peak lit an escape fire in bear grass, then lost it when the winds veered, and one died when he stood up and breathed the searing air. A crew at the Bullion Mine split, the larger party finding its way into a side adit; the rest, eight in all, died in the main shaft. On Setzer Creek, some twenty-eight men, four never identified even as to name, perished as they fled and fought their way uphill and fell in a collapsing ring of death. A gaggle of nineteen spilled off the ridge overlooking Big Creek and sought refuge in the Dittman cabin. When the roof caught fire, they ran out. The first eighteen died where they fell, in a heap along with five horses and two bears; the nineteenth twisted his ankle in crossing the threshold and collapsed to the ground, where he found a sheath of fresh air. Two days later Peter Kinsley, alive, crawled out of a creek. Another group dashed to the Beauchamp cabin, where they met a party of homesteaders. A white pine thundered to the ground and crushed two men immediately, while trapping a third by his ankle; he died, screaming, in the flames. Another seven squirmed into a root cellar, where they roasted alive.

And then there was the crew cobbled together by Ranger Ed Pulaski. He had gone to Wallace for supplies and was returning on the morning of August 20 when the winds picked up their tempo and cast flame before them. He began to meet stragglers, and then a large gang spalled off from the main ridge camp. All in all he gathered forty-five men and with the smoke thickening in stygian darkness turned to race down the ravine of the West Fork toward Wallace. One man lagged and died in the flames. Pulaski hustled the rest over the trail before tucking them into a mine shaft. Then he hurried downcanyon with a wet gunny sack over his head before returning and herding the group into a larger tunnel, the Nicholson adit, which had a seep running through it. Pulaski tried to hold the flames out of the entry timbers and the smoke out of the mine with hatfuls of water and blankets. But by now the men were sense-less. They heard nothing but the din, felt nothing but heat, saw nothing but flame and darkness, smelled only smoke and sweat. As the firestorm swirled by the entrance, someone yelled that he, at least, was getting out. At the entry, he met Ed Pulaski, rudely silhouetted by flames, pistol drawn, threatening to shoot the first man who tried to flee.

By the 1990s, the American fire establishment was a wonder of the world. It could field crews and aircraft to fight fire in numbers larger than the military of some Third World nations. To many critics and to not a few of its own members, it also seemed to have broken. In 1994, wildland fires burned 2.5 million acres of public lands, killed thirty-four firefighters, and swallowed up $965 million off-budget; the summer of 2000 burned still more land, more intensely, and approximately doubled the costs. For all this, a century of federal protection had created a crisis of forest health; many lands suffered either from too much or too little fire, from deluges of wildfire and droughts of fire famine.

The full price of fire suppression became public, along with the admission that firefighting alone could not contain wildfire. But perhaps controlled burning could. That naive formulation finally ended in May 2000 as the National Park Service kindled two prescribed fires under extreme conditions. One escaped Bandelier National Monument and scoured Los Alamos, New Mexico. The other forced the evacuation of the North Rim of the Grand Canyon. It seemed that the American fire establishment could neither adequately fight fires nor light them.

Yet it is possible that the breakdown is not simply one of execution but the upshot of a flawed debate, a false choice between one practice or the other—that we either had to start or had to suppress. But how did that dichotomy happen? Why those choices and no others? The options had become polarized in the usual way, by politics, personalities, and professional pride. In this case, they also had fire to catalyze the social chemistry. With uncanny timing, the polarization happened as the fire crews on the Coeur d'Alene were fighting for their lives.

The idea that fire protection on the public lands meant firefighting was, in 1910, a novelty. Most of the general public was indifferent or hostile to aggressive fire control, bar control of fires that immediately threatened property or lives. Rural Americans relied on fire—burning everything from ditches to fallow fields—and accepted the occasional wildfire as they did floods or tornadoes. The argument that one ought systematically to fight the flames, all of them, seemed odd, academic, and ridiculously expensive. The assumption was that wildfires would go the way of wild animals as the feral landscapes that fed them were domesticated into farms, pastures, and towns. The reservation of extensive lands for public parks and forests, however, broke that laissez-faire logic. Wildland fire would flourish because wildlands prone to fire would persist. In retrospect, the choices are obvious: either convert those lands to something less combustible or do the burning yourself. And that was what critics at the time proposed: abolish the reserves or inoculate the forests by wholesale burning. Better fires of choice than fires of chance.

But in 1910 those options seemed stickier. The national forests existed to preserve the forests, not to wipe them away. If federal agents logged them off, they were no better than lumber companies or homesteaders. If they adopted wholesale burning, they were managing the lands no differently than if the lands had not been reserved at all. To forestry officials, moreover, it appeared plausible that clearing people out of the landscape, fielding patrols, and attacking the wayward flames would be enough. Several decades of "improvements"—roads, trails, telephone lines, lookout towers—would stamp fire out of the scene. This was what the European oracles of forestry argued had to be done and what the great colonial powers were attempting in India, Algeria, Australia, and Africa. It was what Americans had to do.

Yet the critics were adamant. The doctrine of light burning or "the Indian way," as it was called, was remarkably pervasive. Almost all categories of settlers burned and saw no reason to cease. An occasional fire would escape and perhaps raze the occasional town, but that, regrettably, was the price of progress. Smoke in the woods was the complement to smoke from factories. Where land was not farmed but logged or grazed, the preferred means of dampening wildfire was to burn over the understory lightly as often as the fuels would allow. In California, for example, major timber owners hired gangs to prepare sites for burning by filling basal cavities with dirt or raking around snags. They burned after a couple inches of summer rain had fallen. The fires piddled around; they scorched perhaps half the area targeted; they smoldered in windfall. They burned weakly because they had not much to burn. Not every forest could burn this way, but most of those that mattered to people could, and that was how most folk living on the frontier wanted matters. They found intellectuals to back them up, such as the poet Joaquin Miller, the novelist Stewart Edward White, the state engineer of California William Hall, and the Southern Pacific Railroad. Light burning by the American Indian, after all, was what had created the forests for which everyone now lusted.

Yet foresters detested and denounced the practice. However slight its apparent damages, they knew in every fiber of their professional being that it was evil. It sacrificed future growth to current old growth. It abraded soils, gnawed the bases of the big timber, abetted frontier habits of sloth, and promoted folk indifference to the cause of conservation. It was the lost nail that would end with a lost war. To convince the public otherwise—especially those who lived on the land—demanded decades of trench warfare.

But in August 1910, the quarrel took a quantum leap when *Sunset* magazine printed a direct challenge in support of light burning, matching arguments point by point, and even suggesting that the regular army do the burning so that private landowners could protect themselves from federal malfeasance. Then, after the Big Blowup, Secretary of Interior Richard Ballinger championed the cause in a national press release. Clearly, he argued, the Forest Service had failed in its firefight; another strategy was worth pursuing. Government foresters blasted the proposal in the *New York Times*. Fighting fires in forests, Gifford Pinchot enunciated carefully, as though to idiots, was no different than fighting fires in cities. Left unsaid was the corollary: no one would burn off carpets to protect houses from roof fires.

These polar pronouncements placed the Great Fires squarely in the political firestorm that was about to consume the administration of William Howard Taft. Gifford Pinchot had been a favorite of the Roosevelt administration—had free access to TR, had been allowed to trespass across bureaucratic borders, and most critically had convinced the president to transfer the forest reserves from the fumbling General Land Office (GLO) in the Interior Department to the Bureau of Forestry, which Pinchot oversaw for the Agriculture Department. That happened in 1905. Roosevelt also brought in Richard Ballinger to clean up the GLO. Ballinger did, but he quickly posted legal guards along the agency borders and told Pinchot to stay clear. Then Taft arrived as Roosevelt's anointed successor. Like Ballinger, he insisted that Pinchot hold to his own turf, and, worse, he appointed Ballinger as secretary of the interior.

Pinchot found himself, in relative terms, marginalized, and he believed that Taft was similarly marginalizing the grand scheme of Rooseveltian conservation, which for him had assumed the status of a political crusade. When a murky matter involving Ballinger and Alaskan coal lands surfaced, Pinchot seized on it to force Ballinger into disrepute. No formal charges of illegality and corruption were ever fielded, but Pinchot and his allies launched a campaign to discredit Ballinger in the court of public opinion. Someone had to go, Pinchot insisted, and on January 7, 1910, Taft decided that that someone would be Pinchot and fired the chief forester.

Over the next few months, as congressional committees began their own inquiry, Pinchot, the Forest Service, and their allies sought to vindicate the patriarchal Pinchot by vilifying Ballinger. Although both men insisted they were "Rooseveltian conservationists," they represented two very different versions. Pinchot stood for the new wave of technocratic, federal administration; Ballinger, for an old guard, sensitive to local politics and western ambitions. These differences had practical consequences. They mattered, for example, in how each man responded to the problem of fire. For both, the Great Fires became a test of larger philosophies. One had to choose between them. One had either to suppress fires or to start them.

He came to shortly after midnight. No one else in the mineshaft stirred, and at the entrance he found the body of Ed Pulaski crumpled in a heap. He crawled out into a darkness illuminated by flaming snags and logs and began to stagger toward Wallace. There he met the supervisor of the Coeur d'Alene, William Weigle, who had returned from his own misadventures an hour earlier. The eastern third of Wallace was afire. He told Weigle that everyone else at the Nicholson adit was dead.

That was not quite true. Others began rousing themselves, including Big Ed himself. The creek water was hot and alkaline, too foul to drink. They sucked in deep breaths, still heavily laden with smoke. They counted off and realized that five were missing. They had died where they had passed out on the mineshaft floor, probably drowned in the muck and water that had ponded behind the fallen bodies. Pulaski suffered more than most. He was temporarily blind, and his lungs were so charged with soot and seared with heat that he could breath only haltingly. They began the march to Wallace.

As reports screamed across telegraph lines, it was not clear how the fires would be interpreted. Those on the ground considered the Great Firefight as a rout. On the Lolo Forest, supervisor Elers Koch characterized the summer as a "complete failure." More than seventy-eight firefighters had died; the Forest Service had expended almost a million dollars over budget; and the flames had roared over the Bitterroots with no more pause than the Clarks Fork over a boulder. At national headquarters, foresters fretted whether the Great Fires might be the funeral pyre of the besieged Forest Service. In fact, those far removed from the flames saw them otherwise. They chose to see Pulaski's stand, not his flight. They saw a gallant gesture, not an act of desperation. Critics of the Forest Service claimed the Service had been granted ample resources and had failed. Its defenders replied that the Service had failed only because it had not been given enough.

Surprisingly perhaps, the political tide turned in its favor. The Forest Service successfully defended its 1911 budget. The Weeks Act, which would provide for the eastern expansion of the national forests by purchase and for federal-state cooperative programs in forestry, had been stalled for years but broke through the congressional logjam in February. In March, a beleaguered Ballinger asked to resign. Foresters redoubled their efforts to crush light burning, and all it implied. It was, they sniffed, mere "Paiute forestry." Light burners belonged with perpetual motion mechanics and spoon-bending psychics.

The young Forest Service had the memory of the fires spliced into its institutional genes. The Great Fires were the first major crisis faced by Henry Graves, Pinchot's handpicked successor. The next three chief foresters—William Greeley, Robert Stuart, and Ferdinand Augustus Silcox—were all personally on the scene of the fires, had counted its costs, buried its dead, seized upon "smoke in the woods" as their yardstick of progress. Not until this entire generation passed from the scene would the Forest Service consider fire as fit for anything save suppression. Three months after the Big Blowup, Gus Silcox wrote that the lesson of the fires was that they were wholly preventable. All it took was more money, more men, more trails, more political will.

In 1935, Silcox, then chief, had an opportunity to reconsider. The Selway fires of the previous summer had sparked a review in which the Forest Service itself admitted the lands it was protecting at such cost were in worse shape than when the agency had assumed control. Field critics observed that the Service was unable to contain backcountry burning. Scientific critics had announced at the January meeting of the Society of American Foresters that fire was useful and perhaps essential to the silviculture of the longleaf pine. Ed Komarek of the Tall Timbers Research Station observed bitterly that this was the first time such facts had become public. And cultural criticism burst forth as well. Elers Koch noted that the pursuit of fire into the hinterlands—mostly by roads—was destroying some of the cultural value of those lands. The Lolo Pass, through which Lewis and Clark had breached the Rockies, he lamented, was no more, bulldozed into a highway. All this landed on Silcox's desk. His reply was to promulgate, in April, the 10 A.M. Policy, which stipulated as a national goal that every fire would be controlled by ten o'clock the morning following its report. The veteran of 1910 replied that it would be "controlled," that is, by attempting to squash fire, to allow it no sanctuary, to tolerate no qualifications, to apply the full force of the Civilian Conservation Corps and the federal Treasury. He would refight the Great Fires, and this time he would win.

Beneath the surface storms of 1910 politics, moreover, ran a deep current of cultural sentiment. This was an activist age—of political reform, of nation building, of pragmatism as a formal philosophy. One of its mightiest intellects, William James, published his last essay in the same month as the Big Blowup. "The Moral Equivalent of War" argued to redirect the growing militarism James saw boiling over in Western civilization to more constructive purposes. Why, he reasoned, could there not be a moral equivalent of war as there was a mechanical equivalent of heat? Why not divert those martial enthusiasms into a war against humanity's common enemy, the forces of nature, to replace wars against other people?

James had written the essay in Europe. He returned to America, terminally ill, even as Ed Pulaski and William Rock and Joe Halm were standing before flames unlike any short of the Apocalypse. He hurried to his country home in New Hampshire and lay dying as the smoke from the Big Blowup passed overhead and turned the New England sun to a copper disk. The firefight as moral equivalent of war. Why not, indeed?

To an astonishing degree, the Great Fires contain—virtually in all their pieces, voices, and avatars—the grand narrative of American fire. The politics. The contrast between federal and state fire protection. The controversy between fire setting and fire suppressing. The use of the emergency fire fund (1910 saw a twentyfold increase over previous expenditures). The mass hirings and the appeal to the military. Lavish meals in fire camp. Wholesale salvage logging. The (attempted) rehabilitation of burned landscapes. Inaccurate and self-serving reports. The confused memorializing of the dead. The stories of crews saved and crews felled. The metaphor of the firefight as battlefield. A platoon even hauled away an injured bear cub. The profound impact on persons and institutions: the Great Fires acted on the Forest Service as the Long March did on

Red China. Almost every fire story since has its rhetorical structure forged in the flames of 1910, and no fire since has harbored all the parts so completely. Nearly every incident, controversy, or idea had its rehearsal in the Great Fires. Fundamentally the same story plays out at Blackwater, Pepper Run, Mann Gulch, Rattlesnake, Inaja, South Canyon, or the next-millennial fires of 2000. The fires stayed.

Most of the people who fought them, however, did not. They moved on, transferred, climbed up the ranks. A few did hold, Ed Pulaski among them, homesteading bureaucratically in Wallace. At forty, he had been older than most. He was married and had an adopted daughter and a house and remained in the Wallace District until a car accident in 1930 drove him into retirement. Because he stayed, he never left the aftermath of the Great Fires.

He refused to become a celebrity. He wrote only once about his experiences, for an essay contest sponsored by *American Forestry*, and that because he needed money to pay for eye surgery. (Through bureaucratic bumbling, he had received no compensation for his fire injuries.) He tended the two mass graves hastily dug at the Wallace cemetery for those who had died and otherwise were unclaimed and in fact unidentified. He pestered the Forest Service for a more fitting memorial, which finally arrived in 1921. He rebuilt the trails blocked by blowdown; assisted with timber cruising for salvage logging; helped replant the hillsides; upgraded the fire-control organization. When, a decade later, the Missoula office sought to collect the remembered stories of the Great Fires, he declined to contribute. More words didn't matter.

He remained a man who worked in the field and, as all who knew him remarked, a man who took pride in the things he could do with his hands. Practice, not theory, would decide the future. Tools, not ideas, determined what actually could happen. He had led his crew by example, not by exhortation. Acts not texts revealed his meaning. It was fitting then that, after the burns, he devised a combination tool, half ax and half mattock, to send into the field. There was not much enthusiasm when he first presented the device to forest supervisors, but he persisted, lengthening the shaft, widening the ax, shrinking the mattock, all in his backyard forge. He sent the tool out with smokechasers. Only in the field, he insisted, could its value be tested. The smokechasers soon took it to heart. By 1920, so did the Northern Region of the Forest Service, which ordered commercial companies to manufacture it out of industrial steel. Along with the shovel, the pulaski tool became the basic implement of fire control and the one tool both universal and unique to wildland fire.

Contemplate that tool. Three parts make it up: the ax, the hoe, the handle. It's a practical, not an elegant, tool. Cutting and grubbing don't balance easily. It's awkward. It's ungainly. Yet it works, and it embodies the saga of Big Ed and the Big Blowup as nothing else could. Every time a firefighter hefts a pulaski, he or she is retelling the story of the Great Fires.

So, too, an institution like that which governs America's wildland fires requires three things. It needs practice, poetry, and policy. It needs practice to make things happen on the ground. It needs poetry to inspire people, those within its ranks and in the general public both, to make them understand and believe in its purpose. And it needs policy, like a handle, to hold those two opposite-facing edges together.

The Great Fires had it all. They had story, purpose, tools. Modern fire management does not; it holds policy like an empty handle. Fire officers know they cannot continue with the Great Firefight alone. They know fire in the West will return, either wild or by choice; but come it will. If contemporary fire agencies had the chance to replay the controversy over light burning, they would almost certainly choose fire lighting over firefighting as a basis for wildland stewardship. They know the base problem was not fire suppression, but the abolition of controlled burning—that magnificent and misguided attempt at the wholesale exclusion of fire. Yet they know by now that an equally scaled reintroduction isn't possible. The process is not reversible. It can't be undone by simply twisting the other direction, like a screw tightened and then loosened.

The truth is, policy by itself is incompetent. Contemporary fire management has policy aplenty—has had adequate policies for twenty-five to thirty-five years to do what needs to be done without seeing hard results on the ground. Turning existing policy inside out is not likely, alone, to reverse overgrown woods and scrub-infested savannas. Pumping money into controlled fire will do no more good than sluicing money into fire control. Contemporary fire management doesn't need more policy, dumb as an empty handle. It needs a hybrid head of practice and poetry to swing at policy's end.

It needs iron, forge, flame, smith, and vision. It needs knowledge bred into the bone by long practice. It needs flames catalytically equivalent to those of the Great Fires. It needs someone to stand before those flames. It needs a story to explain them. It needs a site that ninety years hence someone can hack into and know that here Creation occurred. It needs a Pulaski.

america's war on fire

"Thou hast conquered a great empire in the saddle. Thou canst not govern it so."
—Ye Liu Chutsai to Genghis Khan

No metaphor has so persisted throughout the history of American fire protection as the firefight as battlefield. Certainly in the early days the metaphor had its substance. Federal protection began when the U.S. cavalry took over Yellowstone National Park in 1886. The occupation of a vast public domain resembled nothing so much as an army of occupation, and the principle needs of their administration, protection against trespass and fire, led to repeated arguments for military troops to be stationed on the forest reserves. And then, mirabile dictu, there occurred that astonishing convergence by which the formative event in U.S. fire history, the 1910 fires, coincided with William James's essay "The Moral Equivalent of War." The extension of fire protection across America's wildlands did indeed mimic a slow wave of conquest, as the fire agencies, like Rome's legions, brought more and more nominally barbaric land under their civilizing administration. Year after year, decade upon decade, the metaphor has glowed and occasionally flamed, as stubborn as smoldering peat. One can count the alternative tropes on the fingers of one hand.

But as Ye Liu Chutsai reminded Genghis Khan, to conquer is not to govern. Whatever power the analogy once held it has shed for journalistic clichés and bureaucratic slogans. It is possible to ask just what it means today and whether it is worth keeping. The metaphor is badly, perhaps fatally flawed, and as a conceptual prism for interpreting wildland fire management it is perhaps even an impediment. The 2000 fire season saw emergency funds for suppression rocket to $1.3 billion, approximately what the United States was committing to Plan Colombia to interdict the flow of cocaine. America's war on fire bore an unsettling resemblance to its war on drugs and for much the same reason: it was not truly a war at all. Wildland fire management clearly needs more than the language of fire suppression.

Why?

- The morality is skewed. War is about people fighting and killing and about
  the coercions necessary to make that happen. It's about people in awful
  circumstances having to make awful choices. It's about moral horror, and
  that, not a bad fire-camp site or a double-lunch tour on a fireline, is
  why war is hell and firefighting is not. Fire control no more shares war's
  moral complexity than does stock-car racing. A hotshot crew has more in
  common with a hockey team than with a combat platoon. Even the gush
  of recent memoirs suggests that firefighting more resembles an extreme
  sport, like white-water kayaking, than a commando raid in the Tora Bora
  Mountains. Those firefighters who die may be mourned, as we would
  a band of alpinists swept away by an avalanche, but they don't deserve
  memorialization as heroes. Nor can William James's proposition work,
  as it once did, as a means to sublimate martial valor into socially useful
  purposes. In Western philosophy, nature is no longer morally neutral.
  Fighting nature has its ethical costs. In practice, firefighting knows that
  dilemma. Firefight-as-metaphor, however, strips it away.
- The metaphor is a cliché, made banal by misuse and overuse. In practice,
  it tends to be shorthand for "we're heroes." Any group is glad to be told
  (or to tell themselves) that they're heroes instead of that, in the end, they're
  only human after all.

     But compare the staleness of that self-congratulation with the power
  of the imagery behind Norman Maclean's *Young Men and Fire*, informed
  ultimately by religious metaphors, or with Karl Weick's analyses of how
  fire organizations behave, with fresh analogies drawn from disciplines
  and models other than military. Which approach brings insight? Which
  brought greater dignity and reconciliation with the tragedies of 1994?
  Which, finally, bestows the greater meaning?
- Besides, military metaphors are many, and their exploitation highly selec-
  tive. Just which military action does a campaign fire mimic? The Battle
  of the Bulge? The Gulf War? Or Vietnam, the invasion of Panama and
  Granada, the policing actions in Somalia, Bosnia, and Haiti? The latter
  would seem to resemble the contemporary land-management dilemma:

How can ecosystem health be restored (i.e., a broken nation be reconstituted) through an external paramilitary force? Probably it can't. But the reluctance to commit firefighters unreservedly after the South Canyon fire in Colorado did not overthrow, by itself, the military metaphor. After all, the military had its own version of South Canyon: the Somalia misadventure that saw army rangers ambushed in Mogadishu. Afterward, neither the military nor the fire community would readily send ground troops if an alternative existed. The fireline-cautious 2000 fire season was thus the equivalent to NATO's Kosovo intervention or the Powell Doctrine as applied to a fiery insurrection in Idaho. In some sense, the military metaphor has persisted, with both firefighting and military strategies morphing in parallel ways.

The compromising analogies, however, are not the ones developed. The cliché describes the putative glories without the complexities and liabilities of the job. It is more cartoon than concept. At some point, the public simply and rightly will become bored.

- Our relationship to fire—and the requirements of fire management—is far too subtle and rich to be subsumed under a military metaphor. The range of what we do with fire far transcends mere firefighting, just as nations are more than their defense establishments. The manipulation of fire is a species monopoly: it is one of the most fundamental attributes to our identity as an ecological being. Yet even as fire management strives to expand beyond paramilitary suppression, no compelling metaphors exist to describe those tasks. Everything is reduced to the same tired, hackneyed phrases of firefight-as-battlefield. The effect is to discard or devalorize everything save suppression. Our metaphors do with language what firefighting emergency monies have done to fire protection overall: they have distorted it to a single, unbalanced purpose.

Words matter. Language shapes, metaphors inform. In fact, wildland fire boasts a chronicle of fantastic experiences and a vernacular full of marvelous expressions—smokechasing, coldtrailing, hotspotting, and so on—that might serve as new metaphors but instead are crushed by the proverbial iron heel of military suppression. If fire management succeeds, it will be not only because new funds are found and new tools and organizations created to reinstate fire in the land, but also because a new language will exist by which to convey the importance and excitement of wildland fire. Suppression conquered an empire; it cannot govern one. It's time for the fire community to get out of the saddle—get off its high horse—and come to ground.

Yet the military metaphor will survive, as it should. Like suppression, it will always have its place and time (though one might wish the imagery and symbolism were freshened and modernized). The critical task is to create new metaphors with which to balance it. The fire community as Sparta is not an enticing vision.

Smokey Bear needs a sibling. Give him a twin brother—call him Flammy—who wields a drip torch. They were separated at birth, but now, arm in arm, shovel and torch, they are together again. So, too, give the military metaphor some complements. Or at least give it a rest.

the
two
cultures
of
fire

The photo is a Rorschach of contemporary fire. The scene has become familiar enough: a house burning amid the woods. That's the background, the deep text of the intermix fire setting, the scrambling of wildland and urban fragments. But it is the foreground that grabs the eye. In what is obviously a scripted pose, a firefighter in full yellow turnout gear is strutting past the camera with the framed photo of a young boy cradled in his arm.

The new American fire frontier has jammed urban and wildland fire agencies together and has forced them to negotiate tactics, tools, even terminology. It also has stirred two very different cultures, the oil and water of urban and wildland firefighting. They are as little bound by shared fire as surfers and irrigation farmers are by common waters. This queasy, immiscible joining the photo reveals perfectly. The setting belongs with wildland fire; the rescue of a youngster, with urban fire services. Wildland fire agencies understand well how to fight fire and poorly how to rescue people. Urban fire services spend less time fighting fire and more saving citizen lives. In this case, plunged into a sparsely inhabited woods, they scrambled to find some point of cultural familiarity. They found it by rescuing a virtual victim.

The urban fire scene is, in principle, as fully under human control as the wildland is under nature's. In principle, people can control everything about urban fire. They can choose incombustible building materials; they can arrange those blocks to frustrate fire spread; they can replace open flame with other ignition sources; they can quickly attack, confine, and extinguish any fiery breakout.

When towns were made of wood and thatch—when they were reconstituted wildlands—they burned like wildland fires. But engineering, building codes, and the wholesale replacement of urban environs by less burnable substances (or previously "fired" materials such as cement, brick, ceramic tile, and asphalt) have strangled flame from the urban landscape. "Learn Not to Burn" is the implacable gospel of urban fire services. Only a fraction of calls that nominal "firefighters" now roll out of the station to answer are, in fact, fires. They are all-purpose emergency crews, their self-identity fixed as lifesavers.

All this makes for an easily constructed moral universe because it means that fire results from breakdowns in the social or ethical order. As the saying goes, the three causes of fire are men, women, and children. The context is one of people behaving foolishly or maliciously or selfishly. Fires track the fissures in the built landscape, breaking through where corrupt politics, fraudulent insurance firms, serial arson, terrorist assaults, or riots strew combustible litter and cast their torches. Fires burn where the oily rags of society have gathered. Urban fire services are among the fiercest moralizers of the modern world. Against society's quagmire of compromisers and phonies stands the firefighter, skilled in his craft, relentless in his selfless quest to protect the innocent. Against the swirl and shoddiness of urban life, the fire service presents a vision of a world properly designed and of a people well behaved.

The story of urban fire is thus a profoundly social narrative. It tells of people interacting with people; the recurring image of the firefighter is that of rescuer, the rugged savior of a helpless victim. If the threatening flames shrivel away, something else can serve the same purpose—bombs, car wrecks, hazardous spills. In novels and personal narratives—the works of, say, Dennis Smith or Larry Brown—the conflict, the moral drama, is an encounter between people, mediated by fire. The firefighters are themselves members of a tightly knit social group, almost monkish, bound not only by training and uniforms but also by unions and an intricate internal hierarchy. The same self-image repeats itself from nineteenth-century Currier & Ives prints to the cover of *America Burning* (the 1974 National Commission on Fire Prevention report) to the Oklahoma City bombing of 1995, where a grieved but unbowed firefighter clutches the bleeding body of a baby. It revived, with firefighters themselves as victims, amid the heroics of September 11, 2001. And if no real victims exist, the genre demands one be invented.

By contrast—an unlikeness almost total—most wildland firefighters do not sign for a career stint, but only for a handful of years. Fire season is a time in their lives, a spell between adolescence and adulthood, when for a few summers they go into the woods, fight fire, and then leave. It is a rite of passage, not a vocation. Its base narrative is a coming-of-age story, which, for all the camaraderie of those times, resides within the individual. Especially in the American West, fires ripple through wildlands, places that are by edict uninhabited public space. The resulting drama is internal: the conflict is to find one's self. The confrontation is between a person and nature; the prevailing group metaphor is the firefight, of an army engaged in a moral equivalent of war. Too often, however, the narrative degenerates into adolescent adventure accounts, juvenile sports stories, and self-promotional posturing.

Fire serves to crank the plot, but because the moral drama typically comes from nature, it remains as thin as flashy grass. A flood or storm can drive the narrative as readily as flame. The ways a person might confront fire are sparser than the infinite interactions a social drama can conjure up. Until very recently, a serious literature of wildland fire scarcely existed. Norman Maclean solved that old quandary by making the firefight into an existentialist drama. His meditation on the Mann Gulch fire captured the profound sense that fire is something transcendent, beyond the ultimate grasp of human control. But he did it so thoroughly that his book may have exhausted what is, after all, a lean genre.

Still, ongoing revelations about fire's ecology suggest other possibilities. What we do to nature is no longer considered ethically neutral. What we do with fire has consequences that both reflect and project human values. In the built environment, combustion can remain a technology, such that the less open flame there is, the better. The ultimate perversion of urban firefighting appears in Ray Bradbury's *Fahrenheit 451*, where firefighters *set* fires and, worse, set flame to books. In natural landscapes, however, combustion can be an ecological process for which there are few surrogates. Our fire practices affect biodiversity, carbon sequestration, nutrient cycling, greenhouse gases. Even more to the point, there exists no neutral position. The era's great urgency is to reinstate fire; the ultimate perversion, to abolish it altogether. The urban imperative to extinguish fire is immutable and unamenable to interpretation. The wildland injunction to restore fire reeks of uncertainty, ambiguity, and irony and has no clear outcome that society can agree upon. In brief, what moral drama wildland fire lacks because of its paltry social circumstances it is acquiring through the ethics of environmentalism.

All this is so alien to the world of urban fire services that one doubts that the two traditions can coexist in the same space. The likeliest scenario is that urban fire will largely drive out wildland, much as industrial combustion is expelling free-burning fire. Local communities will assume responsibility for fire control, create volunteer departments, absorb fire into an urban or exurban protectorate that will ensure not only that houses don't burn but that skies remain clean and that the only good fire is the one that never started. The intermix scene will become, for fire purposes, urban. The majesty of free-burning fire and the glory of our species' ability to start it will retreat to the wilds, where flame can roam untrammeled by the pyrophobia of industrial urbanity and its turnout-coated servants. Fire will retire to landscapes remote in space and surety, where fire service means lighting as well as fighting.

waterdogs

*After heavy thunderstorms, ground moisture often rises as steam and then congeals
into white streamers and puffs of cloud. To the uninitiated, these clouds look for all
the world like smoke. The wary and once-burned know them as* waterdogs. *It often
happens that rookie lookouts report waterdogs as true smokes, in search of which
smokechasers may wander, wet and weary, for long hours through the woods.*

In December, I agreed to lecture at the Environmental Training Centre in
Hinton, Alberta, because I could extend my return through Edmonton for a
couple of days and scout out several caches of documents, primarily the library
annex of the Northern Forestry Centre (NOFC). But I was particularly keen on
the Alberta Provincial Archives, which I was sure would have fire records as
thick as logging slash. In fact, it had almost nothing. There were stacks of
fire reports, just numbers, really; a souvenir guest book from some benighted
lookout who thought he was the next Dharma Bum or Gary Snyder; a few
miscellaneous, probably misfiled reports from municipalities; etc., etc., etc. So
where was the written legacy of the Alberta Forest Service? Everyone I asked told
a different story. In the early 1980s, the ministry had contracted for a history
of its fire program, and I queried the guy who did that mammoth report, a
forestry prof and former dean, now retired, and he shrugged. He had relied
on the published annual reports. My research program, it seemed, had become
laughably simple. I had only to photocopy the hell out of the NOFC reports and
count myself lucky.

Two months later I showed up for a two-week tour, in the dead of a boreal
winter. After several days in the stacks, I thought I would attempt one final
confirmation of the AWOL Alberta archives. Late Friday afternoon, following
several days of phone tag, I finally connected with a guy named Cameron from
the Albert Forest Service, who was doing some history, and he informed me that
the records were there. They just had never made it to the archives. Why, no
one seems sure. One story is that the province simply didn't want to fork over
the money to organize them and go through the drill of formally "archiving."

Another version is that the archives hadn't the storage room and told the agency to house their musty boxes themselves. Neither clutch of procrastinators believed anyone in a hundred years would ever wish to see the stacks (thanks, guys). The fugitive documents are housed, it seems, in the Coronation Records Centre. Think of the concluding scene from *Raiders of the Lost Ark*, a warehouse stuffed with one of everything, its logic understood only by shadowy "top men."

So early Monday I set out. I drove my rental Honda over the North Saskatchewan River to the crowded center of Edmonton, found the skyscraper that houses the ministry, and then miraculously found a parking meter. I headed to the departmental library first because the collections I had been mining had an empty vein for ten years of annual reports, and I hoped to dig out those lost years. That done, I asked the librarian about the Coronation Records Centre. Cam had assured me that I would need only to show up—no permissions, no hassles; he had called all the right people. Then he decamped to Hinton, a dull three-hour drive to the west, no doubt howling all the way, far from the hands that would throttle him. The librarian directed me to an assistant librarian who had previously worked at the Records Centre. No, she insisted, I would need permission. I explained my mission. She wrote everything down, took a business card, and nodded pleasantly. She muttered something about "Willi." So, I finally prodded, who do I talk to about permission? She didn't know. Could she find out? Perhaps; she would check her computer files. Ah, a number. I copied it down. Could I use the phone? No, but there was a public phone at the Information Centre adjacent to the foyer. I hurried downstairs and dashed outside to put more coins in the meter. Then I hustled to the Information Centre, which was vacant. Eventually someone walked out, and I inquired about the phone. Why did I want it? I gave an abbreviated version of the project. . . . I was writing a fire history of Canada, the last fifteen thousand years, but mostly fire institutions, the way people interacted with the land through fire. . . . Her eyebrows knotted in growing suspicion. I then gave the unabridged text, and as her eyes began to glaze, she directed me to the phone. I called. I got an answering machine. I asked the info person if there was someone else I might call. What was the Coronation Records Centre? she wanted to know. I launched into an explanation of where government records went after they left office filing cabinets. Some go

to archives, some to records centers, some to dumpsters, some to people's base-
ments. The eyes glazed, and the hand turned to a directory. Try this number.
I did. Another answering machine. Any other suggestions? When you get the
answering machine, she said, press o to connect to the receptionist. I got a
secretary. She said to come across the street and go to the eighth floor and talk
to the general receptionist. The general receptionist for the eighth floor had no
clue why I was there. Could I explain what I was questing for? Sure. I'm writing
a fire history of Canada. . . . A filmy glaze began to creep across her eyes.
Ah, she finally exclaimed. You need to go to the north tower. This is the south
tower. At the north tower, the eighth-floor receptionist directed me to the library
on the seventh floor. There, the receptionist, between answering phones, took
down the whole saga. American historian; fifteen thousand years; forest fires;
institutions. . . . One of the other staffers, overhearing, then took charge. I
needed to talk to Alice, the departmental records officer. And her number? I
asked, with a gasp. Here. Is there a phone I can use? Yes, at the Information
Centre, across the street. On my way back, I hustle to the meter and drop in
more coins. Alice is out; another answering machine. I press o and connect
to her secretary. When will she be back? Why do I wish to speak with her? I
give her the spiel—I have a compact version by now. She obviously thinks I'm
a crank. Alice won't be available this afternoon. Can I call tomorrow morning?
Let's set a time, I suggest. Okay, 8:30 A.M. At the NOFC, I have no phone. I'm
in an exaggerated broom closet that's used to house all the published literature
no one wants to use. There's a drain in the center of the room; why, I haven't a
clue. But because water doesn't regularly flow down it (the place being stuffed
with books and pamphlets hostile to running water), the drain trap doesn't get
flushed, which means it breeds new life forms and dank odors not entirely
of this world. I have to use the public phone . . . in the lobby. I hurry out
of the north tower and reach my timed-out car a few footsteps ahead of the
meter maid.

I call at 8:30. Alice is not yet in. I'll call in fifteen minutes, I say. By then, Alice
has come and gone to another meeting. She'll be in meetings all day. Alice's
secretary mumbles something that sounds like "Willi." Is there someone else I

can talk to? Otherwise, I'll have to call every fifteen minutes until Alice's and my stars come into alignment. I get another name, Wilma. Wilma is out. I get Wilma's secretary. Wilma's secretary gets the spiel. I'll call back at 11:00 A.M., I declare. I get Wilma. Wilma explains slowly that I will need to get permission. Yes, I say. What I want to know is how to get permission. What does the agency require? Wilma isn't sure. I'll call back at 12:30 P.M., we decide. Then, she still doesn't know. The person she intended to ask is still out to lunch. (They're all out to lunch, but let that pass.) I'll call at 1:15 P.M. In the meantime, the Lanier 6745 photocopier I've rented to scythe through the NOFC library annex, after 7,683 copies, has ground to a halt. I have a service contract. The company agrees they need to send a guy out, but it's across town, so they'd prefer to do it tomorrow morning. Tomorrow morning, I announce, I'll be at the Coronation Records Centre, probably plastering semex around the delivery door. Can they call me back in a little? No, I have no phone. I'll call them at 1:30 P.M. Meanwhile, Wilma has someone for me to talk to. This contact is out. I get a machine. I explain in *Reader's Digest* language what I'm doing. I'll call back at 1:45 P.M. I call InterOffice Machines. They can have someone at NOFC by 8:30 A.M. the next morning. Fine, I say. Wilma has connected me with Karen. I actually reach Karen, who has to confer with someone else. "Willi," she says slowly. "Willi." Can she call me back in a while (like next week)? No, I'll call in half an hour. Now, as Chief Inspector Jacques Clousseau would say, we are getting somewhere. I'll need someone to vouch for me. Who will do? The head of Fire Protection for the Forest Service? The director of NOFC? The deputy minister for Forestry Canada? No, someone at Northern—anyone, apparently—will do. But my prime NOFC contact, Matt, also has decamped for the seminar at Hinton, where he is no doubt sharing all the hilarity of my situation with Cam, over a large mug of Moose Head lager. And the assistant director has jetted off to a conference at Cancún (yeah, right). So I approach Bertie. Hey, Bertie, could you call Karen at this number and tell her I'm doing legitimate scholarship? Yeah, why not? Bertie's a pretty low-echelon munchkin. Playing deputy director has a certain appeal, like dressing up for a masquerade. Karen accepts his call and demands something in writing. He sends her an e-mail message. Apparently that meets specs. Now, I stammer, exhausted with expectation, how do I find anything in the Coronation Records Centre? Oh, Karen assures me, someone from records will contact me and explain. What number should they call? No, I sigh. I'll call you tomorrow, say, 9:00 A.M. You can give me a number then.

The copier gets repaired, but there is no number; there might never be a number. So I elect just to show up at the Records Centre. The security folks are so startled to have an American on their obscure doorstep and no doubt puzzled about what a "fire history" means that they let me in. You should talk to Willi, the receptionist says. Yes, I agree. That's what everyone keeps telling me. Willi, it appears, is on vacation. No one seems to know where, but they agree that he'll be gone for two weeks. He knows everything, I am assured. I talk to his assistant. He listens to the fifteen-millennia saga of Canadian fire, shakes his head, and tells me I should talk to one of the real archivists, Herb. Herb and I have a lovely chat. He's convinced that the records must be somewhere. We have procedures, he explains. There is a chain of authority. But there is nothing here to help tell you how to find anything. We just store and retrieve boxes. We use forklifts, not finding aids. You need to talk to the records-management officer, Alice. No, Willi will know nothing about any of this. He oversees the forklifts. Have a good day.

There is no dignity in chasing a waterdog. You just do it until you know that there is nothing there. At least in the woods, you can collect overtime, and you have the pride of your craft; you can follow even a bogus bearing on a dead line, and you can find your way back out. Scholarly smokechasing is merely demeaning. The secret is, you must have no shame. You simply stumble from encounter to encounter. You play the idiot (that much I have down cold). There is no room for pretense: you know nothing, dignity only costs you time, pride is having your misparked car towed to the police compound. There is no reason, no rhyme, no logic. But if you persist, particularly if you stand there mulishly in its face, the bureaucracy will eventually rouse itself, if only to expel you from its presence until tomorrow. But then, tomorrow is another day.

Note

The names of all the persons involved in this episode have been changed, and in the interest of narrative momentum some incidents and conversations are composites.

a
dumb
problem
to
have

Nineteenth-century America had a spectacularly lethal problem with explosive fires and a stultifyingly simple theory of protection.

The problem was that aggressive westward settlement was mincing forests into coarse kindling and then setting the stack aflame. Some of the fires behaved well enough, much like the pioneering folk who set them. They burned dense woods into stump farms, but a number succumbed to a poorly policed fire frontier and went wild. How to contain the feral few without closing down the westering flood tide, which was not possible in any event, was a challenge.

The common wisdom was that settlement would solve its own problems. Eventually the wildlands would be converted to fields and towns, and the fires would vanish. Conflagrations were a hazard and freak of the frontier like prairie drought or bear maulings. They would pass. Lumber companies even argued ingenuously that they would have to accelerate cutting to stay ahead of the flames, an argument that conveniently ignored the fact that slash from logging and land clearing was what powered the vast burns.

The outcome was a breathtaking sequence of holocausts—Miramichi, Peshtigo, Humboldt, Hinckley, Baudette, Chisholm, Cloquet among them. They were the conflagrating equivalent of border clashes. They would pass, or so it was assumed, as the frontier moved on. And so they did. Both critics and apologists understood that the great burns were only fiery whitecaps whipped by the winds of a vast storm, the westering colonization of America. The fires were sparks thrown by the grindstone, not the force that spun it. They were right that the flames would snuff out as the storm faded and prime movers ran out of steam. When rural America settled into maturity, its fires would be nothing more than the burning of old pasture or autumn leaves.

Rural America, however, did not senesce gracefully. Much of it emptied as its population poured into cities. Large portions of the West never passed beyond the first flush of contact, never knew committed colonists ready to put down roots, before the land was reserved as permanent public domain. Rural fires gradually expired like a guttering candle before the steady onslaught of industrial combustion, housed in machines or propagated over high-tension power lines. Fossil, rather than living, biomass burned. The "fire" problems of this lapsed frontier lay with acrid smoke and coal-fired acid rain. The great burns were a relict memory, less real than the Civil War cannon on the village square.

Then it all came back. Zombielike, the burning of villages revived, the living dead of U.S. fire history. Fires ripped through small towns and suburbs wherever the right mix of drought, wind, fuel, and spark could propel them. Malibu, Oakland, Spokane, Mack Lake, Dude Creek—a half-century after the fires should have vanished, they returned. Yet they revived in a weirdly inverted way, as though rural America had passed through a pyric looking glass. Then and now were uneasy complements. Then, the problem was an excess of fuel caused by cracking open and firing feral lands. Now, the fuel excess results from people *not* cutting, weeding, grazing, planting, pruning, and burning. Then, the crisis was a hunger for arable land driving an influx of peasants, farmers, and herders. Now, the crisis is an equal hunger for recreational land inspiring an influx of city folk. These folk are exurbanites, people with urban ambitions, urban aesthetics, and urban understanding. They do not live off the land, do not depend on a rural economy, but live on it with their money brought from elsewhere. Then, the crisis was considered transitional. Now, the crisis should pass, but only if people choose to imagine a modern equivalent to a rural landscape, a modern, working pastoral. Instead, the wild and the urban—the twin obsessions of American environmentalism—quiver next to each other like matter and antimatter, needing only a spark or jolt to set them off.

Behind the great burns of the nineteenth century lay the deep drivers of
folk colonizing. So, also, behind the flaming orgies of the twentieth, stands
a folkish outpouring. Compare satellite images of Rondônia, Brazil, to Breckenridge, Colorado. The forest of both scissors apart before fractal patterns of roads.
In Rondônia, the land is burned and reburned: this is classic land clearing by
fire, the flames fed by rural migrants. In Breckenridge, the land is emphatically
not burned: it exists, knotted into a thicket of woodlands, precisely because it
has not known regular fire, the fires excluded by exurban migrants. The first has
an excess of abusive fire, the second an absence of obligatory fire. The forced
burning of Amazonian forest is often wrecking the land. The compulsory exclusion of flame from Colorado forests may be no less damaging, for eventually fire
will enter and burn with a ferocity that may scour away not only the housing but
the possibility of regenerating the biota.

The American quandary affects all of its once-rural lands. In the West,
with vast chunks of public land blocked off, this issue properly may be called
an "interface" or wildland/urban interface problem because the private housing
fronts against public wildlands. Elsewhere it results from planting exurban
subdivisions amid a boisterous regrowth on abandoned agricultural fields. Or
it may result, as in Oakland's Berkeley Hills, when the city imports wildlands
in the form of close-packed trees within its boundaries. A general expression
to cover all these conditions is to call it an "intermix" fire scene. Or consider
it simply an ecological omelette. We have leaped from a rural frying pan into
an exurban fire.

But whatever its nomenclature, the fires of this New Frontier, like those
of the Old, will persist until the conditions that sustain them disappear. There
are many regions—the Northeast, for example—that are unlikely to suffer huge
wildfires because the fundamental conditions for fire spread do not exist. There
is no regular rhythm of wetting and drying and no addictive usage of fire
to wrest a rural scene into habitable shape. But where the land is fire prone
and particularly where it is exposed to dry lightning, an outbreak of intermix
burning will flourish as long as that peculiar frontier persists. It will burn and
reburn in Florida and California and Montana and Arizona until the crush of
colonizers converts the scene completely into an enclave of urban fire protection

or until the frontier collapses and the land reverts to the wild or until some unexpected condominium of good sense prevails and the frontier reaches an equilibrium in which a working replacement of a rural landscape emerges. Although no doubt shaped by urban values, this replacement will be a lived-in landscape, one that finds people as active agents in shaping the fire regime of their surroundings, not simply passive victims and whining litigants.

There is no reason to have houses burn on this scale. This is a stupid crisis because it is a fixable problem. Eventually the press of colonization will spend itself, like a wave that washes up on a beach, broken. Until then, the fire community has limited power to halt the flames. It cannot stand in the surf and fatuously command the colonizing tide to turn with any more hope of success than King Knut, but it can help keep structures from burning and prevent a heedless loss of lives. We have replaced the flames of rural hearth with a microwave oven. We next need to replace the frying pan.

when
the
mountains
roared,
again

The fires of 2000 were big, expensive, in our face. But controlling them—or waiting until nature shut them down—may have been the easy part. The real firefight is deciphering their meaning.

For that, one needs context and comparison. We know things not in themselves but in their relationship to other things; this is as true for fire as for anything else. But especially with fire, context is all because fire has no identity apart from the spark, wind, fuel, and terrain that sustain it. Fire synthesizes its surroundings. We know it by what it catalyzes. Its context is as much cultural as natural, as stocked with symbols, signifiers, institutions, and information as with pine needles and snags. So by what social metrics do we understand the 2000 fires?

The simplest standard is sheer size. With 6.9 million acres burned, thirty thousand firefighters called out, and fire costs overall likely to top $2.5 billion, 2000 was a millennial year. Yet size by itself says little. No one remembers the supposedly epochal 1951 season against which 2000 has measured itself; nothing dates as quickly as costs; and Ethiopia mobilized seventy thousand citizens to battle the catastrophic fires in its highlands, about which no one outside the country had a clue. Even in America, bigness is no guarantee of importance. Large fires savaged Washington in 1902 and the Adirondacks in 1903, but memory of them passed through the culture like smoke. They endure only as regional relics.

Then there is politics. For a month, wildfires were the darling of the media, and in an election year they were bound to prod politicians in a country flush with revenue to drop money like fire retardant. Accordingly, the federal agencies were primed to receive $1.8 billion in new funds for fire protection.

How deeply the fires have passed into the belly of the American political beast, however, is unclear. What Congress gives, Congress can take away. Those monies had a specific target: to eliminate the spectacle of burning houses and, not incidentally, to provide jobs to rural communities. The fires disappeared from media screens with the first rains and next month's disaster story. The intermix fire is only one of the West's many fire problems. Nothing is more common than an ashbed effect after a bad fire season, flush with fiscal fireweeds. And nowhere in the scramble to establish either legitimacy or legacy after the controversial presidential election was there mention of the fires of 2000.

The political ecology of fire, in fact, much resembles slash-and-burn cultivation. For a season, it is possible to plant in the ash. Wait two years, however, and the weeds of everyday life crowd out the cultigens. Social memory overgrows; the media moves feverishly on to new disturbances. If the burning is combined with some other practice—logging, grazing, sowing—its effects can be extended. So a big fire can leverage its influence if it connects with larger movements. Still, scale and timing matter. If the fires are a part of a general wreckage, their urgency may be lost. They must tap into major political movements, but not too large. Too little shock, and the fires may be safely ignored or shrugged off as a climatic aberration. Too much shock—say, if accompanied by political crisis as in Indonesia—and there may be little will or institutional apparatus by which to deal with them. Burning malls in Jakarta will mean more than fired fields in East Kalimantan.

In U.S. history, prominent fire reforms have accompanied eras of major reform of many institutions. Large fires acted as catalysts in what was, for other reasons, a favorable context for action. The 1910 fires occurred during a time of peace and relative prosperity that was also a time recognized as a period of political activism and reform (the Progressive Era). No other crisis obsessed the nation, save a furious controversy over conservation policy, to which the fires spoke with violent eloquence. The Great Fires could command center stage.

Another era of restructuring occurred during the Great Depression, in which large fires in the backcountry coincided with single-party domination of both Congress and the presidency (a coincidence rare in U.S. history) and with a president determined to merge conservation programs with public works. Policy debates and eventually changes during the 1960s and 1970s surfaced amid a de facto social revolution. Probably it is no accident that the federal agencies did not complete their reforms until after the Vietnam War and the Watergate crisis had concluded. The push for a common fire policy among federal agencies and for better rules of engagement for committing firefighters that followed the 1994 season resulted not only because thirty-four firefighters died that summer, but because a best-selling book, *Young Man and Fire*, created a prism for public appreciation and, not least, because the Republicans seized a majority in the House of Representatives for the first time in half a century and were rabid to shake up the federal government. A fire can act only on what exists around it. In this case, the cultural landscape was as primed for a blowup as the brushy slopes of Storm King Mountain. The National Fire Plan that emerged in late 2000 promises to be the most comprehensive review since the 1933 Copeland Report. It builds on a small mountain of studies, from institutions as diverse as the General Accounting Office to the National Academy of Public Administration, no less than the usual suspects—the federal land agencies. It is unclear, however, whether the National Fire Plan will find a New Deal on whose shoulders it can ride or whether after September 11, 2001, fire protection may be cradled into an agenda of homeland security, a sort of interior Coast Guard.

All these episodes suggest that fire reform depends on timing: it requires a general crisis, highlighted by fire, sufficient to scare the political establishment but not so damaging that it cripples the capacity to act. The fires have to defibrillate a stunned body politic. If the body politic has had only the breath knocked out of it, it will ignore the fire after it regains its air. If the larger shock, however, has killed the body politic, then the fires are simply kicking a cadaver to no purpose.

Perhaps the 2000 fires were a distinctive kind of fire, the pilot flame for a new era. It is, in fact, possible to parse America's fire history as a succession of "problem" fires, each of which has in its turn commanded inordinate money and staffing, a cycle that beats to a twenty-year rhythm. (One can trace this odd phenomenon back to the creation of the earliest forest reserves by presidential proclamation in 1891.) The latest avatar—the bedeviling question of how to cope with a recolonization of rural America by exurbanites, a process that has created an ecological omelette of houses and scrub—emerged roughly in 1990. Certainly the spectacle of fires engulfing houses was a prominent feature of the 2000 season and the one that grabbed Congress by its lapels.

But placing the 2000 fires into the intermix stew paradoxically diminishes their claim to uniqueness. They simply confirm for Idaho and Montana what has become commonplace in California and Florida. They offer a newer version of what everyone already recognizes. They become one of a queue, roughly at the midpoint of an era. That is not a claim to uniqueness. The money that an election-year Congress lavished on the problem came with marching orders, very explicit expectations, which if not met could cause the funding to dry up. Congressional attention may disappear with the rains or a meager season and with the media's rush to the next disaster story. From afar, much of the local-based hiring may look like rural pork, and the beefed-up investment in fire science may resemble a form of political money laundering.

Yet another rhythm exists in U.S. fire history, and upon it one might claim a special character for the 2000 season. This rhythm beats to a long-swell, thirty-year story cycle. (Fire history has no single orbit. Instead, it resembles a solar system, full of planets and comets and stray asteroids, all in their own periodicities and pathways.) The contours of that narrative rhythm look like this.

They begin with the Transfer Act of 1905, which brought the forest reserves to the U.S. Forest Service, followed by the Great Fires of 1910, the ur-fire of modern American experience. These fires wrote the originating narrative that would guide national understanding of wildland fire throughout the rest of the century. They were great not because they were big, although big they certainly were (five million acres on the national forests alone), but because they burst out of the political fissure between Gifford Pinchot, chief of the Forest

Service, and Richard Ballinger, secretary of the interior, a rift that would help
sunder the administration of William Howard Taft. They ignited the first Great
Firefight of the Forest Service. Before it ended, seventy-eight firefighters had
died, the Forest Service had roared a million dollars over budget, and the agency
institutionalized the trauma through a succession of chief foresters, each on
the firelines, that continued until 1939. The Great Fires also sparked a national
debate over fire policy, the controversy over light burning. And they captured
a national thirst for a "life of strenuous endeavor," as Teddy Roosevelt put it,
or for what William James urged as a "moral equivalent of war." Firefighting
embodied such sentiments exactly, so that in intellectual circles, no less than in
the field, firefighting attached itself to the prevailing culture. All these threads
the Great Fires knotted together into a master narrative—a story of fire control,
the firefight as the basis for our relationship to wildland fire—that guided the
Forest Service and through it the other federal agencies as well as the Service's
state cooperators until the early 1930s.

By 1935, the story faced a crisis. The droughts of the early 1930s powered
large fires in the backcountry. Especially the 1934 fires that savaged the Selways
forced the Forest Service into a soul-searching review of what it had accom-
plished or had failed to accomplish since 1910. So, too, light burning resurfaced
in the form of southern critics who argued for the value (even the necessity)
for fire in the management of the southern pines, for grazing, for wildlife, for
fuel abatement, for control of bluespot fungus. The politics was also favorable:
the Roosevelt administration was eager to pair the rehabilitation of America's
Depression economy with that of its wrecked land. The Civilian Conservation
Corps, in particular, placed the means to rework fire practices into the hands
of those who had the power to decide ends. Thirty years after the Transfer Act,
Chief Forester Gus Silcox, a veteran of the 1910 fires, promulgated the 10 A.M.
Policy, announcing a program of all-out fire control as an "experiment on a
continent scale." This was a more intense and lavish narrative of fire protection,
though one that shot out of the rifled narrative bore of the Great Fires.

Thirty years later, the story underwent another revision. The narrative
weave had frayed badly. The economics of fire control had reached a point of
diminishing returns. The ecology of fire exclusion made clear that the removal
of fire had costs, that no true neutrality existed with regard to fire; that in

wildlands prone to fire some kind of fire would exist; that the choice was between wildfire and controlled burning. An efflorescence in ecological science confirmed a vital role for fire in many biotas. The precipitating factors, however, were cultural, best symbolized by the Wilderness Act of 1964. A simple firefight was no longer an adequate strategy for fire in natural areas. Critics urged some attempt at "restoring" fire to its pre-Columbian state, at least in complement, if not in full-bodied substitution.

The question of an adequate story, however, was less certain. In 1967–68, the National Park Service reformed its fire policy to encourage more fire, preferably natural fire left to burn under its own circumstances, but controlled burning where that was necessary. In 1978–79, the Forest Service followed suit. The era became one of wide experimentation, under a growing conviction that because fire was "natural," it was also good.

The prescribed fire story, however, tended to grow in parallel to the fire-suppression story and, in fact, in defiance to it. Prescribed fire's narrative never fully replaced suppression's and, in striking ways, never established itself on its own terms. It remained indebted to fire suppression intellectually, no less than infrastructurally. It grew around the suppression narrative like a milpa's bean vine around a maize stalk. The philosophy behind the fire potlatch at Yellowstone in 1988 was as garbled as its implementation was opulent. Not least, prescribed fire had its failures—breakdowns, escapes, an inability to get sufficient acres burned on the ground or burned in the right way. When stressed enough, the master narrative defaulted to the firefight, which defined the template that prescribed fire had inverted and which remained the action of ultimate power and last resort. Controlled burning rested, finally, on the ability to control fire.

Advance another thirty years, from 1964 to 1994. The 1994 season had a catalytic effect akin to that of 1934, and the reform of federal policy that followed in 1995 was a modern avatar of the reforms of 1935. What failed to emerge, however, was a clear narrative that contained something of the elemental power that the 1910 fires had bestowed. Suppression clearly had little juice left to squeeze from its pulp, just the specter of fallen firefighters (though they more resembled trekkers lost on Everest than defenders of the national estate). But prescribed fire—in a dialectic that had taken shape first in 1910—offered an inadequate alternative. There was no catalytic event comparable to the Great Fires.

Until, perhaps, 2000. The fires have passed; their context continues to shift. It's too early to say what their lingering impact will be, though not too early to hope. The events may hold latent meaning, released like seeds freed by flame from the serotinous cones of lodgepole pine and ready to sprout with the next rains. It may be that the analogy frequently made between the fires of 2000 and the Great Fires of 1910 holds, that the 2000 season's fires may act on the reforms of 1995 as the big burn of 1910 did on the formative events of 1905. It may be that a new narrative will emerge from the ash that transcends the dichotomies that have confined the story so far and that have so spectacularly revealed their flaws.

This would be the ideal outcome: the fires of 2000 join the pantheon of Olympian blazes. Jolted by the new-millennial fires, America seizes the chance to reform. The National Fire Plan, the Joint Fire Science Program, a flurry of general publicity—these become the basis for a fundamental reconstitution of how America relates to fire. But it just as easily may be the case that the flames will blow away like the Florida hanging chads of the 2000 election. Either way, either choice, the outcome makes equally great history. But the two outcomes do not make an equally great present. The nation needs reform on the ground.

The systemic breakdowns of 2000—the failure to control prescribed fire at Bandelier National Monument, the inability to halt wildfire in the Rockies— beg for something more, a narrative that might redefine the means and ends of fire management. The future does not reside with either of the two camps of fire fundamentalists or with some bogus middle ground of quasi-natural, quasi-prescribed, quasi-suppressed burning. The proper focus for debate over fire resides with the character of the fires' setting. In fact, the season did hear, with dramatic clarity, some new voices on just this theme. The core problem, some claimed, was not housing where it shouldn't be, but the overgrown countryside within which those structures resided. The crisis lay with forests clogged with overstocked and undernourished trees that invited disease and insects, smothered landscape patchiness and species diversity, so that fires that formerly rippled through surface bunchgrasses were shunted into canopies, where they soared like a storm surge. The cause of the enormous burning was not simply drought but a legacy of how we have related to the land. The summer's kindling was the sad legacy of the mismanagement of fire over the past 150 years. This was the fuel behind the fury. The spark of the summer's dry lightning was only as powerful as the combustibles it fed upon.

All this, however, supposes the fires will not fade with the lost smokes from snag and stump and with the crimped flow of dollars from the federal treasury. It supposes that a collective consciousness can work through the char and confusion and construct a social story out of the wreckage. Yet common stories have not become all that common in contemporary America. More likely is a rash of personal cosmologies, springing up like fireweeds, a population of individuals peering like solipsistic shamans to divine personal meaning in the play of flame and smoke.

Even that may be too great an expectation of intellection. The memory of the fires ultimately may reside where they began, with those who lived in the cloying smoke of the Bitterroot Valley, those who watched fires in sheets like torn sails blown by tempests, those who stood under the shadow of smoke columns as black as raven wings, those who trod over incinerated hillsides, their once-living biomass blasted to biotic bits. These fires would join the other magnificent fire busts of the Northern Rockies like those of 1919, 1929, 1934, 1941, and the rest of a long legacy.

Great fires, however, demand more. If the fires of 2000 are to become great fires and not simply remain big fires, they must spark more than shiny new air tankers, engines, and drip torches. The environmental tragedy of fire in America is not that wildfires were suppressed but that controlled burns ceased to be set. The cultural tragedy is not that post–World War II fire specialists were technically incompetent but that they were no longer literate in the great philosophical and historical themes of significance to their society. They increasingly knew what ought to be done. They just didn't know how to tell everyone else.

burning
off

burning
deserts

The second thing you notice is the eerie similarity to fires everywhere. The flames crackle through stiff grasses, surge up trunks, flash through tangled brush. Air tankers drop red-retardant slurry, helicopters circle like hawks, and crews clad in hard hats and yellow nomex cut, dig, and burn out. Wildfire and backfire collide.

But that resemblance strikes you only after you pass through the unsettling realization that what is burning is the Sonoran Desert. Not the clichéd metaphors of burning sands, flaming sunsets, fiery or furnace-hot deserts, but the real thing: a Sonoran Desert stocked with grasses, ocotillos, ironwood, chollas, and saguaros. The firefight is genuine. The two fires, metaphoric and real, converge.

The fact is, Arizona is built to burn. Fire is no more alien to the desert than is water. The winter rains and summer downpours, the soaking drizzles and flash floods, are not only an analogue for fire but the source of its fuels. Under natural conditions, free-burning fire is a two-cycle engine of wet and dry conditions, sparked by lightning. Without rain, there would be no fire. Fire's regimes mimic those of water, which is why, in Arizona, something is nearly always burning. In the mountains, fires follow droughts. The deep-stored waters in soil and trees drain away; fuels that are typically too wet become available for burning. In the desert, fires trail wet winters in which steady rains sprout dormant seeds of grasses and forbs and carpet rocky soils with wildflowers and fine fuels that dry quickly during the long-arid spring and carry fire as easily as tissue paper. The marvelous biotic mix that characterizes Arizona assures that some fraction of the state is almost always ready to combust.

Weather determines when. The Southwest's celebrated thunderstorms—dry on one side of a road, wet on the other—are ideal for separating lightning from quenching rain. Early summer storms in particular drip sheets of precipitation, virga, that evaporate away as they descend; lightning and winds slam the ground, but rain vanishes, a marvelous formula equally for dust storms and firestorms. Winter rain, spring drought, summer monsoon—the rhythm is perfect for burning, and fires rise and fall throughout the year with those seasonal tides. On a longer rhythm, they move up and down slope with the climatic storm surges of El Niño and La Niña.

Its fabled terrain shapes both local storm and life zone strata. The Mogollon Rim piles up seasonal thunderheads like spume off a seawall. The sky islands of the Basin and Range—the Chiricahuas, the Peloncillos, the Bradshaws, and all the others—draw isolated thunderheads like high islands in the Pacific. The sudden drop of canyons, ravines, and rims helps segregate dry spark from wet rain, allowing fire to forage for suitable fuels. Those complex contours create niches for combustion. All they need is a spark.

But nature did not have to kindle all of Arizona's fires and in all likelihood did not do so. People kept fire constantly on the land. In fire-prone Arizona, whoever controlled ignition controlled fire and through fire the larger environment. Humans competed with lightning to determine what regime would shape the regional ecology. What people first burned, lightning could not. The fire regimes became those of hunters, foragers, irrigation agriculturalists, transients, and warriors.

They burned as people always have. They fired, deliberately or accidentally, along their corridors of travel. They torched patches in which they wanted to hunt, drawing deer, rabbits, and elk to the fresh growth of grasses and shrubs that sprouted from the ash. They fired around mesquite, piñon pine, and oak to assist the gathering of beans, nuts, and acorns. They burned around habitations to prevent ambush and wildfire. Sometimes their burning smoldered through wet or feeble fuels; sometimes it sailed wildly before drought and high winds. As they cycled seasonally through the landscape, they carried fire with them. Their lines of fire and fields of fire were the stitches and patches that created a landscape quilt. Their escaped fires became the stray threads and floppy yarn that loosened ties and let odd pieces flap.

Overall they encouraged a more open landscape, one flushed with fine fuels, stocked with scattered large trees, dappled with scattered shrubs and rich browse, abundant with frequent fires. As the menagerie of Pleistocene megafauna disappeared, they left still more fodder for the flames. Few of the giants survived, yet one, a fire creature, became dominant by working with a fire-prone climate to sculpt whole landscapes. Much as Arizona's indigenous peoples manipulated water's flow by building brush dams and diverting streams and relying on irrigation rather than storms, so they reconstructed the region's fire regimes.

European colonization shocked this system, both by what settlers did and by what they did not do. The first jolt was the removal of most native burners. The "missing fires" that would become so prominent a century later were not simply those that nature set and people suppressed but those that people had once set and no longer did. That left as an ignition source only lightning, and it was lumpy, producing fires that more closely mimicked the erratic ebbs and flows of the regional climate. The second shock was the introduction of livestock on an immense scale. Sheep and cattle cropped off the grasses that had traditionally carried fire, lightly but regularly, through forest, chaparral, savanna, and even deserts episodically flushed with fuels. Logging for mines and railroads locally broke down the structure of old-growth groves that had stood amid prairielike understories. Then the permanent reservation of public lands, particularly for national forests, imposed active fire suppression as an agency mandate.

Settlement shattered the alliance between human firebrands and grasses that had so shaped aboriginal Arizona. Woody floras that previously had proved unable to survive the spring firefloods took root, flourished, and overgrew every landscape from desert playa to alpine krummholz. Ponderosa forests often knotted into dog-hair thickets of sickly poles; juniper and mesquite spread over former prairies like gangrene; chaparral closed, replacing its supple texture with a stuccolike overcoat of shrubs; desert grasslands morphed into cacti and brush, while exotic trees probed outward from riparian refugia. The denser fuels and more erratic ignition meant that fire regimes shifted from one of frequent, light fires to one of more infrequent, intense fires. A landscape that had simmered chronically now began to boil over.

Probably the Sonoran Desert shared this scenario. What affected every other Arizona landscape likely affected it as well. A desert that had known swift-passing surface fires perhaps three or four times a century—fires that scarred saguaros but left them standing, as mountain fires did ponderosa pine—now sprouted alien weeds like brome that served as fuses to carry fire to the thickening brush and there detonate. These fires did not cull; they killed. In the desert, as elsewhere, the number of large fires in all biomes has increased steadily over the past decades. Wildfire has replaced controlled fire.

More surprising, perhaps, is the realization that the greatest shock has not come from more fire but less. To strip fire away from a place, for example, that has known surface burns every three to four years and to replace it with a conflagration every forty is akin to ending regular winter rains and concentrating a year's precipitation into a single summer thunderstorm. The system will wobble and likely crack.

How to restore fire is not obvious, easy, or uncontroversial, particularly when industrial fire, especially internal combustion, drives Arizona's contemporary ecology. Urban values, an urbanized population, urban fire-protection models—these determine the perception and response to fire. In metropolitan areas such as Phoenix and Tucson, there is not airshed enough for both free-burning fire and automobile exhaust. The Lone fire of 1996 that engulfed the Four Peaks wilderness and sent enough smoke to shut down Sky Harbor Airport temporarily was a dramatic illustration of what is being played out daily in the grand transition from open flame to internal combustion.

Still, Arizona's wildlands remain. Far more than its golf-coursed metropoli, they are an emblem of what makes the state special. As public lands, they are not inhabited; they will not be converted to tile-roofed stucco suburbs or shopping malls; they will burn. If humans refuse to oversee the burning, lightning fire will move in, like coyotes into vacant lots. The same conditions that create Arizona's flaming sunsets will continue to put real fire on its mountains.

Fire, like the water with which it is linked, is both good and bad. Without water, the desert dies, and human societies will dry up. But the desert requires a particular regime, and humans need a controlled flow, not a flood. So it is also with fire. In some environments, fire is necessary; in most, it is simply present, a part of what has made that landscape what it is. Contemporary Arizona will no more abolish fire than it will mountain or monsoon. What both land and society need, however, are controlled burns, not wildfires.

This transition will be difficult, as immense as the construction of Hoover and Roosevelt Dams, as meticulous as the lacework of canals that yet grid the Phoenix metropolitan area. Fire-restoration projects have sprouted principally in the state's wilderness areas—on Nature Conservancy sites such as Gray's Ranch, the Malpais Borderland Group in the southeast, and along the slopes of Mount Trumbull—all places where fire is ecologically indispensable and where smoke is remote from cities. The amount of fire restored is a fraction of what existed when the first American fur trappers probed down the Gila; it is unlikely to expand to anything like the former dimensions of free-burning fire.

So the firefights will continue, as they must, even along desert-fringed golf courses, along the roads that join the urban and rural lands, and amid city parklands. In 1995, a lightning-kindled fire swept 80 percent of McDowell Mountains Regional Park. The episode reminds us that fire will find its way, that Arizona needs to think as deeply about its fire regimes as it has about its reservoirs and rivers. Fire, like water, there will be: the mix of wild and controlled is yet to be determined.

a
story
to
tell

Anyone even casually acquainted with America's wildland fire scene knows the chasm between fire control and fire use. Fire control holds the money; fire suppression created and maintains the institutional infrastructure; and firefighting historically has dominated the culture of wildland fire management. Until recent decades, no one stood accused of misbehavior for suppressing smokes, as they might be held liable for a kindled flame that escaped or, more tellingly, for failing to burn a site that craved it. The assumption was that one fought a fire unless told not to but set a fire only after every conceivable contingency and clientele was satisfied. Whatever public policy urged and personal philosophy prompted, the reality was a powerful bias in favor of fire control.

It is natural, then, that proponents of controlled burning should try to correct that imbalance by matching suppression, item by item. That they should demand flexible funding, and lots of it, similar to the emergency fire accounts. That they should field burn squads analogous to hotshot crews. That they should create a parallel program of certification. That they should seek to change liability laws to create some legal space for burning and to tweak environmental edicts to accommodate smoke. That they should host National Prescribed Fire Awards akin to the Golden Smokeys. That they should found a National Prescribed Fire Academy akin to fire suppression's National Advanced Resources Technology Center. That they should conduct fire setting on the model of firefighting, complete with similar language, tools, and elan.

They may succeed. The old landscapes, however, did not result from a regimen of burning modeled on suppression, so it is doubtful that this particular process will recover exactly what fire exclusion has lost. But it doesn't have to: almost any fire in such sites may be better than none at all. The deeper issue is what it will take to slash through all the institutional scrub and burn away public skepticism. The fact is, suppression is a false rival. Controlled fire does not face fire control like two bull elks bugling a challenge and locking antlers, one or the other to triumph. Rather, it sinks from the bites of a million mosquitoes, reddened into frustration, plagued into lethargy. Suppression is not, in truth, the problem. Controlled fire must make its own case, not rise out of the ruins of fire control.

For this, it needs a story. Criticism leads to skepticism; story, to action. The real, the most elemental difference between fire control and fire use is that firefighting can tell a marvelous story, an environmental epic, but prescribed burning cannot. It is easy to forget that fire control did not take the country by storm. From the beginning, it fought a bitter, decades-long policy battle with proponents of light burning, and it laid an even more stubborn siege with public opinion, one whose citadel did not crumble until after World War II. Until then, the larger American public was indifferent or hostile to wildland fire control. Fire suppression faced every bit as many challenges as controlled burning does today. Yet it overcame them all.

The process took decades, luck, and bureaucratic grit. Why did proponents of fire control persist? Why did the public finally believe them and not their rivals? The simplest explanation is that they had a powerful story to tell and their critics did not. As nearly as one can date such matters, that narrative emerged out of the wreckage of the 1910 conflagrations. Majestic, huge, lethal, the Big Blowup became the founding saga—a Kalevala, a Song of Roland—for wildland fire's heroic age. The narrative of 1910 explained what firefighting meant, and it got institutionalized to the virtual exclusion of any other narrative. To it, America owes its wildland fire establishment. Light burners had no such saga to sing. Neither do advocates of prescribed fire today.

Until such a story appears, it is doubtful that controlled burning will succeed to the extent that its advocates desire and America's wildlands deserve. It isn't enough for controlled fire to continue to swat mosquitoes, even in the millions. It needs the capacity to ignore them, to bull ahead through the muskeg of politics and public opinion, confident that it will thrive in the end. Nor is it enough to downgrade suppression. Fire control's loss is not necessarily controlled fire's gain. The problem is not suppression, which is necessary, or the literary set-piece saga of the firefight, but the absence of a complementary story for controlled burning. Prescribed fire does not need more policy. It needs a poet.

compassing about with sparks

Behold, all ye that kindle a fire, that compass
yourselves about with sparks: walk in the
light of your fire, and in the sparks that
ye have kindled. This shall ye have of mine
hand; ye shall lie down in sorrow.

—Isaiah 50:11

The crisis boiled over in 1935. The previous summer's fires in the Northern Rockies had sparked a board of review that questioned whether fire protection in the backcountry was possible or, to the mind of one dissenter, even desirable. The January meeting of the Society of American Foresters had sponsored a session, under the direction of Yale professor H. H. Chapman, its president, on the value of controlled burning along the southern coastal plain. The papers, which argued for fire, were a revelation. That spring, with competing claims before him, Chief Forester F. A. Silcox promulgated the 10 A.M. Policy as a universal standard for fire suppression, stipulating that every fire would be controlled by ten o'clock the morning following its report or, failing that, by ten o'clock the day after that. Questions of whether fire suppression was the right strategy or whether it was truly possible were brushed aside. Even critics characterized themselves as "heretics."

In truth, professional critics were few. Wildland fire control had fought a bitter battle to establish itself before a skeptical public. The Forest Service had begun fire protection only in 1905 or, more effectively, after the Great Fires of 1910. As a bureaucratic exercise, it had thrived under heavy pressure and with mixed results for only twenty-five to thirty years. Its supporters believed that the public could accept only a simple, unified message—that fire was bad and its suppression good. Anything less would gut fire protection's grander mission. Enthusiasts explained away continued breakdowns by the policy's failure to enjoy the political support it deserved and the money it required. The 10 A.M. Policy gave suppression virtual carte blanche; the Roosevelt administration lavished men and money on it, and professional forestry drowned out the voices of doubters.

There are echoes of an eerie symmetry between then and now. Prescribed burning commenced as a formal policy in 1967–68 with the National Park Service; the Forest Service modified policies until, in 1978–79, it adopted a parallel program; a common federal-agency policy emerged in December 1995 with a charge to handle fire in an "appropriate" way, a mix of firefighting and fire lighting. Depending on when one chooses to date its origins, the doctrine of prescribed fire has lasted roughly as long as the doctrine of systematic fire protection before the 1935 crisis. Prescribed burning is not a new program. Its breakdowns—and they have been many, even lethal ones—no longer can be explained away by bureaucratic stupor, public hostility, or the need to build up a repository of experience. The time has come to ask whether the existing regime of prescribed burning is, in fact, fundamentally flawed or whether it requires only a fuller commitment, a kind of Clintonesque health care program for wildlands. That the National Park Service lost two widely separated prescribed fires, both set under extreme conditions in spring 2000, recalls the passage from *The Importance of Being Earnest* in which Aunt Augusta observes to the orphaned Ernest that to lose one parent may be regarded as a misfortune, but to lose both sounds like carelessness. In this sense, we have returned to 1935, with the escaped Cerro Grande and Outlet fires taking the place of the Pete King–Selway fires of 1934.

Now the eerie part. Where is the board of review, questioning whether prescribed fire should expand as suppression did? Where are the equivalents of the Society of American Foresters' session on controlled burning? Where are the professional skeptics, suggesting that prescribed burning may not be a universal solution? The fire community, rather, seems determined to defend prescribed fire as an ideal, despite its lapses; to continue to believe that the public cannot accept any kind of nuanced message; and to behave as though any doubt cast onto the value and technical possibilities of prescribed burning will kill an irreplaceable program. Instead, they present controlled burning as obligatory and hence beyond discussion. Often advocates describe it as easier, cheaper, and safer than suppression. It is, inherently, none of these.

The fire community has presented the public with a false dichotomy: either fire suppression, which has failed, or prescribed fire, which continues to limp along only because it has not been granted sufficient support. We must light fires because we cannot fight them. The only way to protect Los Alamos from burning down is to risk burning it down accidentally. Fire is natural, and fuels are excessive; there is no other option. Nonsense. As in 1935, the fire community seems to fear that, confronted with a compromised message, the public will throw the good out with the bad. Yet that is, in fact, what will happen if bad burns are justified as good.

Prescribed burning has racked up an impressive litany of failures. Ouzel, Mack Lake, Seney, Gallagher Peak, Pocket, Yellowstone, Lowden Ranch—these are among the best known. But escapes have occurred yearly. Official statistics often hide more than they reveal. Over the past twenty years, the worst fires of several entire seasons have been prescribed fires gone bad. Since 1979, seven prescribed fires have killed firefighters (a new twist on the concept of friendly fire). Smoke flooding across roads has caused fatal car accidents. In 1980, an escaped fire burned Mack Lake, Michigan, to the ground. The two most costly firefights of U.S. history were prescribed burns that blew up. Because escaped fires become reclassified as wildfires, such failures are not always reckoned in final tallies. These breakdowns will continue, and if prescribed burning scales up, so will their size and cost. The statistics on escaped fires are notoriously unreliable, but a consensus estimate suggests that 2 to 3 percent escape. This is, uncannily, the same rate of failure experienced by suppression.

The more serious failures, however, are those fires that simply don't do the job they should. They burn too cool, too hot, too long, too spottily, too infrequently. They create as much dead fuel as they remove. They occur as an isolated burn, not as part of a suite of practices and a series of sustained fires. More broadly, the program has not racked up the necessary acreage. Controlled fire simply isn't happening on the order promised or required. Much of the burning is occurring not where it is most needed (the West) but where it most easily can be done (the South). Some acreage that was registered formally under "wildfire" is now registered as "wildland fire use." Areas are burned to meet targets, but not necessarily burned right, just as they were once logged willy-nilly to "get out the cut." And then there are the promised fires that are never lit. None enter formally into the running register of failures.

Why has prescribed fire not flourished? The reasons are legion. Liability law, smoke, threats to endangered species and cultural resources, complicated land ownership, narrow "windows" for burning, insecure expertise, competing purposes, lands overstuffed with fuels—take your pick. Returning fire to a landscape that has not known it for five to ten decades is tricky. Preparation (thinning, mostly) is expensive and often arouses public ire. Fires in heavy duff or logs will smolder, perhaps for weeks. The fire community has hardly begun to address issues of greenhouse gases (has barely begun to explain why, given global injunctions to sequester carbon, they find it necessary to uproot and burn off tens of millions of acres laden with carbon-stuffed biomass). Controlled burning is reemerging out of institutions and funding designed to extinguish fire, which is a structurally different task than igniting them. Wildfires create crises, which must be addressed; unlit prescribed burns pass unnoticed. The list goes on.

But the nuclear reason may be the absence of a truly compelling reason to burn, one that engages the public and links fire to culture as well as to ecology. The problem is not one of policy but of poetry, a passionate conviction of the heart, ideally expressed in a story that can inspire as well as inform. Controlled fire has instead two narrower convictions—that fire is natural and that wildland fuels are excessive. Both are true, and both irrelevant. The fire-is-natural argument calms critics worried that the proposed meddling is something that the lands cannot tolerate. It notes that fire has been on the planet for four hundred million years and that many biotas are as adapted to particular fire regimes as they are to patterns of rainfall. Since the mid-1960s, the "restoration" of fire has enjoyed the same cachet as the reintroduction of wolves—a necessary cog refitted into broken ecological machinery, a powerful symbol of wild America, and penance for past environmental wrongs. The preferred means was to let natural fires free-burn over wildlands in the form of "prescribed natural fires." This was the program under which, at a cost of more than $130 million, Yellowstone National Park burned 945,000 acres, although Yellowstone didn't bother with the nuisance of prescriptions. Unsurprisingly, the term is no longer used.

The more serious critique is that the fires to which the western landscapes had adapted formerly were not strictly "natural." They were, rather, the result of complicated interactions between nature and people. Nature contributed rhythms of wetting and drying, grew fuels and readied them for burning, and broadcast lightning over select landscapes. But humans were there to push their own ignitions, to rearrange fuels en masse, and to carry fire where lightning couldn't. American Indians burned widely, and so did frontiersmen, and they created an elastic matrix within which lightning fire had to operate. This had gone on since the end of the Wisconsin Glaciation. It ended in the West only when overgrazing cropped off the grasses that had carried the flames and when settlement removed the indigenes who had started them. Creating national parks and forests confirmed the trend toward fire exclusion by making fire protection a deliberate policy. Granted this history, it is difficult to know what *natural* means or why it justifies prescribed burning. The ancient chronicle of human burning, however, does argue for people putting fire back in.

The more useful plea or, rather, threat—the alarm that has pried money out of Congress—is that a history of excluding fire has stockpiled the public wildlands with surplus fuels to the point that a fire, from any source, virtually can detonate. This is unquestionably true in places, and high-intensity fires beyond the historic range of burning are becoming normal, again in select places. But it is not true everywhere. Woods clogged with combustibles do not erupt spontaneously into flame, and there are many long-unburned sites in temperate lands that become less fire prone with the passing years. Piling pine needles does not add to fire hazard because only the upper crust will carry the flaming front. Adding annual rings to old-growth Douglas fir does nothing to worsen fire hazards and in fact lessens these hazards by enlarging a heat sink. In places without regular wet-dry cycles, deciduous forests may crowd out conifers and smother fire in the shade. It certainly does not follow that restoring fire is the only means to check conflagrations. The exclusion of fire, alone, did not create the current crisis, and the reintroduction of fire, unaided, will not correct it.

Fire works best in nature as it does in the lab, as a catalyst. It interacts. It quickens, shakes, forces. What caused the fuel buildup was not simply the forced absence of fire but the linked changes between fire and human land use—grazing, logging, hunting, farming, foraging. Fire by itself cannot reverse those massive shifts, and thrusting a torch into fuel-choked sites today, without suitable preparations, is an incitement to ecological riot. Fire is not ecological pixie dust that, sprinkled over degraded lands, magically renders them hale. Even if the fire does not escape, it may burn under conditions far different from those of the past. The places that most need fire are often those whose fuels cannot now readily accept a spark. It is not possible to flash-burn a forest the way oil wells can flare off unwanted gas. Too many foresters, in particular, continue to conceive of fire as a "tool"; to imagine it as a mechanical force for removing, thinning, shuffling, harvesting; to imagine that it behaves like well-trained foresters rather than as a biological presence whose power derives from its setting.

The argument that only prescribed fire can reduce fuel is absurd. If excess fuel is the problem, then remove it. Haul it off, burn it in fireplaces and powerplants, mulch it into compost, send it through woodchippers. Crush it, crop it. Browse it with goats. Thin and stack it before burning. Burn sun-dried cuttings while the surrounding woods are still green. Burn piles in the snow. On sites dense with combustibles, a burn may yield more dead fuel than it consumes. A fire may sputter and smolder, unable to gnaw through thick veneers of woody matter. Prescribed fire, in brief, is not a miraculous cure that, on touch, dissolves away the leprosy of woody litter. Some experimental fires have succeeded in slow-cooking large trees and sparing small ones, exactly the reverse of historic patterns and intended outcomes. Besides, a prescribed burn is not a vaccination, a one-off inoculation against conflagration. It typically involves a series of burns, often with complex preparations, then repeated in perpetuity. The burns presumably become easier as they successively return, and once a site has plumped into such a state, regular burning is a marvelous means to keep fuels under wraps. But that must be part of a twenty- to thirty-year program.

Prescribed fire requires a better justification. For burning to be mandatory, worth almost any risk, the critical consideration is not reducing fuel but promoting the biotic cycling that fire sets into motion. No alternative technology exists because free-burning fire is not so much a tool as a captured ecological process, less a flaming ax than a dancing grizzly. Combustion works biologically in ways that chain saws and woodchippers don't. We have it backward. We don't need fire to reduce fuel; we need fuel to allow fire to work its ecological magic. After all, "fuel" is not carbon bullion: it is the product of living organisms. Our determination of ecological needs should help us select the kind of fire we want, which will determine the kind of fuel necessary.

The outcome will be mixed, as it should be. There will be some places where fire is mandatory, some where it is useful and optional, some where it is irrelevant or simply dangerous. The outcome should not be merely a fuel-reduced landscape, but a fire regime suitable for the biota and, where people reside, a habitable landscape.

Purposes cannot be separated from practices. Part of why controlled burning has faltered may lie in how we do it. The federal agencies conduct prescribed fires as they do firefights. The legal and regulatory environments in which flame must exist today push agencies in this direction, and so does their own history. Controlled burning is reemerging out of institutions designed to fight fire. This is awkward, and the National Park Service in particular has tried to counter it by creating parallel, mirror-image organizations.

The scheme is understandable, and mad. It ignores the fact that the two tasks depend on each other—fire suppression requires burning, and prescribed fire requires control. Their divorce can lead to situations like that at Upper Frijoles Canyon in which the burn boss in charge of ignition is not certified to oversee suppression should the flames escape. Equally, it dismisses the fundamental differentness of the tasks, that fighting and lighting should operate through divergent styles and methods on the ground. Suppression looks much the same everywhere because it is a reaction. Prescribed fire should look different with every site. A peril of prescribed fire is that it will simply invert suppression, in both principle and practice; that it will repeat with the left hand the failures of the right. To conduct prescribed fire on a fire-suppression model is, in the end, to share its costs, risks, and dangers. Any single factor may shut the project down, although none allows it to jump ahead. Scheduled burns are a formula for constant attrition. The program will slip year after year.

Instead, we need a constellation of fire practices. We need to spot burn, to sequence burn, to prescribe crown burns. On larger landscapes, fire managers may need to be fire foragers—constantly searching out small niches of fuels, as a bear might seek out huckleberry patches; prowling the snowline with piddling burns; constantly moving and burning, not in one grand set piece of fire kindling but in a finicky gathering of fuel and flame. Such a program requires lots of fire, lots of smoke, lots of time in the field. It means that most places can't accept such a regimen because they have changed too profoundly since the forced removal of fire. It means that urban critics will object to the chronic smoke and wilderness critics to the fact that people, not nature, are setting the fires. It means that the reintroduction of fire will not succeed in many places, that fire's domain will be a scant fraction of what it was a century and a half ago, that it will cluster especially in remote places. This may be the best we can do.

Be sad. Many of us believe the western landscapes need more fire of the right sort. Some of us consider the proper manipulation of fire, a species monopoly, as a species obligation. We shouldn't turn the task back to nature because nature gave it to us. (If we really want to appreciate the awfulness of bungled fire, wait until nature grants the power to another species.) But if the right fire is the right thing for the land, then we have to find a way to do it and not pretend that bad fires are good. We can't allow feral fire to roam through towns like rabid coyotes. Rogue burns such as the Cerro Grande and Outlet deserve to be hunted down and shot.

Failures of omission, failures of commission—all of them could be excused and were excused when prescribed fire was fresh and experience sparse. The practice is no longer new, and if thirty years is not a long enough learning curve, then one might question on what basis we should expand programs and push them into new lands. Thirty years was enough for critics of fire suppression to recognize its limits. Yet the suppression crowd might shout back that it had never been given the tools—the men, the money, the machines—it needed to do the job right. If the choice were between continuing the firefight or reverting to light burning, they wanted the fight. They would never turn back to the presuppression era.

Now it is prescribed fire's turn. How long will it take before critics rise up and declare that prescribed burning has failed to live up to its promise? Most fire officials came of age during the controversy over free-burning fire in the wilderness; too many seem locked into the Good Fight against the Smokey-Bear-kill-every-fire era, insisting that fire is natural, that there can be no return to the old ways, that prescribed burning has never received the money and staffing it needs to do the job right. This misses the point precisely. The antagonist is no longer fire suppression. The debate is internal, over the right ends, means, and places to reinstate fire. The choice is not between fighting fires and lighting them but over the proper ways and times to do each, within a context that transcends either practice alone. The danger is that programmatic momentum simply may roll on and that, as in 1935, there may be a tendency among Old Believers full of conviction and frustration to replace on-the-ground facts with political fiat and agency "targets" for burning, thinning, expending, and staffing that may or may not have much relevance in the duff.

As the 2000 fire season migrated from the hydrophobic prescribed burns of the Southwest to the wildfires of the Northern Rockies, the spreading confla-grations deflected the early-season critique of prescribed fire, rather as disap-pearing hard drives from the vaults of the Los Alamos National Laboratory diverted attention from the scandal of careless burning to the scandal of sloven security. The shift proved useful, much as media misrepresentation in the early stages of the 1988 Yellowstone fires proved ironically helpful, because in both cases apologists could direct attention away from practice to policy and away from policy to philosophy. The argument fixated on the question of whether fire had any legitimacy in the Jemez Mountains and whether prescribed burning in the abstract had any justification as a device of fire management. The answer to both was, "Of course." The unwise exclusion of fire had helped create the hazard. The capacity to start fires was an essential complement to the ability to stop them. Such challenges the National Park Service knew it could fend off.

But these were not the core issues. The matter under dispute was whether the fire program at Bandelier made sense—whether kindling fires under such circumstances was warranted and, beyond that, how it was that a federal agency could devise a project so flawed and then believe it could talk its way out of patent failure. Alarmed that the public might choose to cancel prescribed fire altogether, the fire community, as it had at Yellowstone, again started to circle its engines. Initial reviews were highly critical; subsequent boards of inquiry were concerned, as one might expect, to protect a policy that a generation had labored for all their professional careers to promote. They had to safeguard prescribed burning. The failure at Upper Frijoles Canyon lay in minor matters of execution; the breakdown occurred because fire suppression had too long held sway, the land was too starved for fire, and prescribed burning still did not enjoy the full confidence of the political establishment. In contrast, Secretary of Interior Bruce Babbitt, who had done more to promote fire programs than any previous secretary, instantly appreciated the magnitude of the crisis and shrewdly moved to admit guilt, to place the Park Service prescribed fire program under a moratorium, and to agree to pay $661 million in compensation to the residents of Los Alamos. This was not a matter for the courts.

It was, nonetheless, a matter for the court of public opinion, and that discussion never had its full day of hearings. The 2000 fire season, roaring through the Rockies, relocated discourse back to the limited power of suppression. The National Fire Plan, released a month before the presidential election, promised to pour money into rehabilitation and the protection of houses, thus redirecting debate away from burning to thinning. The assumption endured that fire would return.

Probably it would. As long as we have wildlands in fire-prone climates, which is the indisputable status of the West, we will have fires. The option still exists to choose what kind we can live with. Probably, too, as further judgment weighs in and the mounting costs become apparent, the realm of prescribed fire will continue to shrink, even as the philosophy behind it billows in the winds of biocentric enthusiasms. It may be that the American public is prepared to accept booms and busts in nature's economy, however extravagant, as it accepts those in the national economy—that the 2000 season will prove to be the fire equivalent of the looming stock market crash, with Cerro Grande a fitting emblem for the Nasdaq Nineties, an overhyped prescribedfire.com.

The 2000 fire season was perhaps a missed opportunity for the fire community to take the full complexity of its charge before the public. What was missed was the chance for a full airing not merely of the policies of prescribed fire but of its praxis. If prescribed burning is more than metaphysical hype, it is, as any true American pragmatist will tell you, only as good as its doing. Its risks are real. Often they are worth taking. The public deserves a clear explanation of why and when that is true, and it needs to hear, with unblinking clarity, that Upper Frijoles Canyon was not such a time and place.

an
incident
at
praxis

Star Trek: The Undiscovered Country *opens with the accidental explosion that nearly obliterates the Klingon moon Praxis. With flames rising behind him, a Klingon spokesman assures Captain Sulu that "there has been an incident at Praxis" but that everything is under control. There is no reason to violate treaty stipulations. There is no cause for assistance. In fact, everything has changed.*

The prescribed burn that escaped Upper Frijoles Canyon in Bandelier National Monument and blasted across forty-eight thousand acres, 260 structures, and a chunk of the Los Alamos National Laboratory in May 2000 with cleanup costs in excess of $661 million is a study in misdirection. That the fire went on a bearing it was not intended to go is not a surprise. The real surprise is that anyone should be surprised and that critics have not seized on the fundamental cultural factors that drove the fire's ignition as fully as wind and drought powered its free-burning flames.

In such circumstances, irony comes easily. The misdirections were many.

One: Bandelier is a national monument established to protect Anasazi ruins, yet it committed to a wholesale fire-management program that had little to do with the relic stones. This was calculated mission creep that became mission misdirection.

Two: The burn's announced objective was primarily to reduce fuel and hence fire hazard. Upper Frijoles Canyon holds some of the densest vegetation in Bandelier. But biomass is not fuel: it becomes fuel only if it exists in a form available for burning. Upper Frijoles Canyon held a lot of vegetation, not a lot of fuel. It harbored exceptional biomass because it was wet—which is to say, because it was not disposed to burn. A prescribed fire in 1992 failed to ignite.

Accordingly, the April 2000 "prescription" called for extreme conditions. Why a woods that could burn only, perhaps, under extreme conditions warranted a hazard-reduction burn is not obvious. In the end, having scorched virtually everything in sight, the worst regional wildfire in memory burned Upper Frijoles Canyon itself only patchily. There's no fuel like an old fuel.

Three: The assertion that burning was the best means to reduce fuel is based on the dual beliefs that fire is a tool and hence a legitimate method of treatment, and that fire is natural and thereby better than "artificial" tools. These claims are philosophical, and they have the effect of privileging fire. Certainly flame predates people, but there is no evidence that the fire regimes of Bandelier were solely the outcome of natural processes or that deliberate kindling out of the "natural" fire season must produce "natural" results. The likely conclusion is that under purely natural conditions Upper Frijoles Canyon would almost never burn.

Moreover, if fire is a tool, then it should compete with other tools. If fuel reduction is the objective, then mechanical thinning is often a more reasonable method than burning. In fact, the most successful burns in Bandelier (along the rims) were done in association with thinning. Again, the argument put forward that fuel reduction was the burn's purpose is at odds with the techniques used. Actual practice suggests that the point was to "restore" fire and that heavy "fuel" accumulations were used to justify that end.

Four: The largest pot of money to do burning in the national parks explictly targets fuel reduction. Although there may be good ecological reasons for reinstating fire, a program must appeal to fuel reduction to access those funds. This distorts programs into doing fuel reduction, whether warranted or not, as a device to get money to do what most fire officers probably regard as the more significant need, the reintroduction of fire as an ecological presence. The fear of large fires stoked by excessive fuels has frightened the public and Congress into allocating funds, though not necessarily into programs that those on the ground would most wish for. Thus, such programs are often based on a fundamental misdirection: they say one thing and do another.

Five: The fuel-reduction argument trots out as its ultimate justification the threat to Los Alamos posed by wildfire in Bandelier. The lightning-caused La Mesa fire in 1977, in truth, did burn onto laboratory grounds. A wildfire in

1996 again threatened to spread into the town and lab. This is obvious. With the prevailing winds from the southwest, a schoolchild could predict the vector of a fire's spread. But if the risk lies at Los Alamos, then why target Bandelier as a solution? Between the monument and the town runs extensive forest, prone to crown fire. An escaped fire at Bandelier will go where all other fires at this time of year go: to the town. The more obvious strategy would be to work from the town outward. But this would have stalled a fire program at Bandelier because it could not have justified money to do burning in the name of fuel reduction.

Six: Park Service apologists have sought to deflect criticism away from the prescribed burn onto the escaped fire. Their reasoning is this: the prescribed fire did escape but could probably have been contained with a rapid response by knowledgeable fire crews. Instead, the task fell to less-experienced personnel who kindled a backfire along a nearby road. This backfire is what blasted out of Upper Frijoles Canyon and raced across the Pajarito Plateau. Such arguments are jesuitical at best. There would have been no escaped backfire without an escaped prescribed fire. That adequate fire-suppression forces were not on hand is a flaw within the burn plan. All other agencies in the region had declined to burn because of the extreme conditions and, in fact, had asked the monument not to proceed.

Seven: The assumption was that the landscape (and fire regimes) that existed in presettlement times were better than those of the present day. By almost any measure, most thoughtful observers would agree. The Pajarito Plateau in, say, the mid–nineteenth century was more open and grassy, less prone to explosive crown fires; its pines more vigorous and less susceptible to insects, disease, and parasites; its surface vegetation more diverse; its overall appearance more inviting. The restoration of fire, it is implied, would restore such scenes.

Probably they might, but not by themselves. The exclusion of fire began with intensive grazing. Sheep are "natural," yet flocks in huge numbers can closely crop land in "unnatural" ways. Simply removing sheep will not mean the land goes back to what existed before they arrived. So it is with fire, and that is why the simple fire-is-natural assertion is insufficient. The issue is not with whether fire might belong: of course it does. The real issue is how it should exist, in what regimes, by what methods, at what costs.

Here appears a historical misdirection. The time warp at Bandelier was not between the closing nineteenth and opening twenty-first centuries but between the present and the 1960s, when the Park Service formulated a new philosophy of fire. This was, at that time, a heroic creation. It flew in the face of public expectation and against the pressures of a bureaucratic juggernaut that sought fire out with the ruthlessness of an antiterrorist brigade. The Park Service not only argued that fire belonged, but also called the existing system of aggressive fire control a threat to biocentric values. Firefighting, not fire, was the intrusion. Fire was natural, fire suppression unnatural. Where natural fire could not roam wild and free, prescribed fire could be accepted as an honorable if less-pristine substitute.

There the National Park Service has remained. The Old Believers continue to staff its programs. The same philosophy continues to guide management decisions. When confronted with apparent failures, the agency turns to its touchstone theses: fire is natural; the problem is a legacy of fire suppression; fire must be restored. When Yellowstone erupted in 1988, the official defense was that crown fires were a natural part of the Yellowstone biota. Critics might reply that, yes, this was true, but the question had moved beyond such assertions. The issue was not whether fire in some capacity belonged in Yellowstone but how the park should go about managing it. That second discussion never flew. So when the Cerro Grande fire blew up, the National Park Service could reply, as it had since the 1960s, only that fire was natural, that they had done nothing wrong, that suppression (that old bogey) not prescribed fire had caused the breakdown, that fire was so essential that the American public must tolerate some escapes. They continued to criticize a "Smokey Bear mentality" when the public had long recognized that Smokey's message went to children and was ready for a more adult rendering.

The big issues—the metaphysics of fire, if you will—had progressed over the course of thirty-five years. Contemporary *philosophes* of fire tried to sort out how natural and anthropogenic fire together had shaped historic landscapes. They tried to explore how fire—as an interactive technology, as an ecological catalyst—worked with other processes to shape biotas. They were ready to explicate *how* fire belonged, not whether it ought to or not; the National Park Service was not. It was committed to burning. It remained fixed in a majestic disputation from the mid-1960s; this time, however, it had the money and staffing to force its ideas through instead of merely discourse about them at national conferences. And so the agency did. At the same time, it lost not only Upper Frijoles Canyon Number 2 at Bandelier, but also the prescribed Outlet fire on the North Rim of Grand Canyon, which forced an evacuation and burned unimpeded until it expired, gasping, on the rim itself. Those paired fires should propel us beyond irony to tragedy and to a conceptual catharsis.

The charge against the agency is not simply that a willful fire went in a direction other than that a willful fire program had planned. The concern is more systemic. The fires resulted not from a series of miscalculations but a sequence of calculated misdirections. The explosion at Bandelier likely will prove more consequential in shaping future fire programs than the 2000 season firefight in Idaho and Montana. This is no time to recirculate hoary fire fundamentalisms, like papal encyclicals appealing to Augustine and Aquinas for their authority. This is less about metaphysics than about pulaskis in the duff. The fact is, fire is messy. Fire is ambiguous. Fire frustrates prescriptions. Fire resists easy control. Its meaning is found not in axioms but where torch, needles, and shovels meet. There is no philosophy without praxis.

doc
smith's
history
lesson

He walks with the aid of a fancy cane now, and his voice has lost some of its rasp. His wireless glasses give him a vaguely donnish look. The black-metal pole looks like something between an increment borer and the sleek stalk of a sotol. He uses it as a pointer, though his blackboard may be a panorama thirty miles wide and twenty deep. It has something of the power of a swagger stick, without the swagger, of which he has no need. Anyone who meets him instantly recognizes he still has some bark on him. He doesn't mutter anymore about unhelpful critics, that they "chaff my ass." He speaks instead about what most engages him, the woods. Specifically, he holds forth, in a gentle manner and with the conviction born of long experience, about the ponderosa pine forests of northern Arizona. They are a mess. Doc Smith knows why, and he thinks he knows what to do about it.

He pulls out an increment borer from his pack and hands it to a youngster to finger and puzzle over. He is standing amid a pine patch on the Coconino National Forest outside Flagstaff, Arizona. He shows the boy how to grip the borer and then points the boy and the tool at a ponderosa pine approximately twenty inches in diameter and tells him to twist. The borer spirals inward. The boy repositions his feet, and Doc tells him to be patient, as he works the borer another turn. When he has gone halfway, Doc tells him to stop and reverse his twists. The borer creaks and screams and slowly spins out. From it, Doc extracts a lengthy core of wood, the size and shape of a long drinking straw. The group crowds around. The core shows clearly the tree's annual rings. He lets the boy begin counting them. This dowel of wood, he explains, is a historical document.

It contains the pith of the past. That's where our understanding must begin, Doc says.

He walks to a large pine nearby. He points to the base, where a big cavity draws their attention (a "cat-face," he calls it) and describes how surface fires have hollowed out that cavity. After the fire, the tree puts on new growth and buries the scar within. Then another fire arrives, and the tree absorbs that scar. This is what the end result looks like, he tells them, as he plucks out of his pack another object, a plank of wood like a thick, warped boomerang, the cross-section of a tree. Its surface is planed smooth and buffed and shellacked. The scars show clearly. Labels date them, year after year. The tree that "donated" this section, he chuckles, came from a short distance away.

This slab, he explains, is a hard-text chronicle of fires for this place. It's a minimum number because not every fire will enter the cat-face. Note, he emphasizes, holding the slab high, how often fires occurred until here—he points to a label titled "1876"—and that no fires scarred the tree before it died a century later. This is typical of trees and landscapes everywhere in the Southwest. Lots of fires until the late nineteenth century, few fires since then. The big pines you see around here grew up under the old regime. The small guys grew up later. Many of them grew up all at once, in 1919. (Early foresters in the region, he noted, almost muttering, wondered how they could coax ponderosa into better reproduction—what it took to get thick regeneration in this climate. Then came 1919, and they found out.)

That's the simple story, he concludes, as he heads back to the van. You can't understand what the problems are and what to do about them until you know how they came about. It's called history.

Doc's hand sweeps across the horizon. He is standing on the sunny south-facing side of a picnic area that looks to the San Francisco Peaks. Before them stretches a park, a dry swamp of bunchgrasses that at a distance resembles a prairie. To the southeast rises Mount Humphreys, a dormant volcano and the tallest peak in Arizona. To the south pushes up a somewhat smaller peak, Mount Kendricks. Their slopes are clothed with forest—pine from the parks up the flanks, then often aspen, then spruce and fir. The vast swath of pine is grey and black and barren of needles, as though a great flood had scoured through the woods, splashing and plucking in a violent, cascading rush. The floods were, in fact, wildfires. The Hochdoeffer and Horseshoe fires of 1996, the Pumpkin fire of 2000, a host of lesser burns: like a San Andreas fault, rupturing at its oldest stretches as the strain builds, a sequence of crown fires has been sweeping through the plateau's woods, part of the largest contiguous ponderosa pine forest in the world, and stripping them clean. Another twenty years of such burns and the entire panorama will be scorched, seared, or perhaps simply swept away.

This is not how fires used to burn here, Doc tells the group. There have always been plenty of fires. The Southwest reeks of lightning fire. Year after year the area gets pounded. The years with the most burning were those that turned dry after several years of abnormal wetting, a rhythm linked to the beat of El Niño. But although fires were common, they stayed mostly on the ground. The pines grew large and in clumps, and the landscape resembled a pine steppe, a temperate savanna. The fires burned briskly through bunchgrasses. The forest was open: the frequent burns, every three to six years, sometimes more often, flushed out the understory and prevented serious pine reproduction, much as spring floods scour out their channels. There were exceptions, of course. The land is not a simple mesa; it holds ravines and even canyons and cinder cones, and good seed years fluffed into patches of dense thickets that could surge into flame. But all in all the land burned widely and often, not intensely. A surface fire might torch an occasional tree, but such fires did not propagate crown fires, not in the pine belt.

It's different today. More and more of the ponderosa forests vanish into violent fires that do not slosh along the ground but gut the canopies and everything in between. Decade by decade the number of large fires has increased, and they are large not merely by their size but by their furor. So today nasty, all-sweeping crown fires are savaging the plateau. Doc holds up his fire-scarred slab. We changed regimes, he explains. The little fires went; the big fires have replaced them.

Doc knows well what this means. He spent thirty-five years in the Forest Service, all of it in some capacity in fire, and has a fresco of plaques, like multiple diplomas, on his office wall to testify to his achievements. He moved up the fire ranks with the big fires. He first achieved "fire boss" status in 1974. Next he became a Type I incident commander in Region Two, the central Rockies. In 1986, he served as area commander for complexes of large fires in Alaska, with multiple incident commanders under him. He helped teach the "fire generalship" course at the national level. Then it was California during the Siege of '87, Yellowstone in '88, and a rash of outbreaks in the Boise National Forest in '89 and '92. After he retired, he was asked to return, like Cincinnatus, to assume area command over yet another eruption of fire through the Boise pine forests that were experiencing high-intensity flame unlike anything their evolutionary history had prepared them to accept.

These were the same kind of fires that have raked over the Coconino: cognates to those burns that extend beyond his evocative hand. Doc Smith has seen them elsewhere because the same crisis has afflicted most of the western woods. He has watched such fires firsthand, has understood the futility of trying to fight them *mano a mano,* knows that the Great Firefight, although magnificent and drenched in adrenaline, cannot continue. The fact that we tried to fight them all is, paradoxically, part of the problem. More history, he mutters. There has to be a better way.

The experiments at Chimney Springs began in the 1970s. Doc tromps through the dead, fine-stemmed pine, in places tangled like a briar patch. The idea was simple enough: Since the exclusion of fire was a primary cause for the overgrown, undernourished woods, its once-profuse glades of forbs and grasses now paved over with pine needles, the solution was to reverse that trend. Put fire back in. After a fire or two, under the caressing hands of prescribed burning, the forest would correct itself. The small pine saplings would burn away, the needles would incinerate and leave flowers and fungi to thrive in its ash, the large old-growth ponderosa—the West's fabled yellow pine—would recover vitality and again be inoculated against crown fires.

It didn't happen. Fire simply synthesized its surroundings: the dense, overgrown woods, groaning under its load of duff, gave rise to equally problematic fires. These fires killed the young growth but could not burn it up, which left them to stand and shed needles, adding to the overall fuel load. They smoldered in the knee-deep duff under the old-growth, slow-cooking roots that had grown up into the organic soil created by the long-unburned accumulation of decomposing needles, and, after a year or two, the great giants often died. The end result could be fuels as serious as those that originally prompted the experiment. In a few places, controlled fire worked its alchemy. In most, it only catalyzed whatever was there. Fire alone was insufficient because fire does indeed take its character from its context, and that context, as Doc patiently points out, was historical. It took him a long time to come to that realization.

A small-brimmed Stetson covers his balding head. He stoops a bit, lowering his five feet, ten inches of height, but the chunky torso still testifies to a youth of hard labor split between Mississippi, where he was born and spent World War II, and La Luz, New Mexico, where he grew up, and El Paso, Texas, where he graduated from high school in 1954 before heading into the navy. It was while he was in the navy that he got his nickname; he can't recall why. "No big story behind it—it just seemed like a good name." After his tours at sea, he turned to the woods, and that meant fire. He saw his first fire in 1958 as a smokechaser on the Wallace District in northern Idaho, ground zero of the Great Fires of 1910. He impressed Don Durland, his boss, who in 1949 had been a smokejumper, the year of the Mann Gulch fire. The next season Durland wangled Doc into the Missoula jumpers. These were the glory years of aerial fire protection. Prior to World War II, the federal government had bulked up fire protection, almost overnight, through the Civilian Conservation Corps. A full recovery waited until the end of the Korean War, when surplus military hardware sloshed into civilian uses, particularly through the Forest Service and its fire cooperators. "Air superiority" became a shibboleth in America's cold war on fire. It was all around him: the skies roared with engines observing smokes, carrying firefighters, dropping chemical retardants, resupplying remote fire camps. At Missoula, Doc found he had joined a fraternity that claimed for itself the pinnacle of the era's ambition. All in all, it was a powerful elixir on a young man: this was where the action was.

As with so many future rangers, fire was Doc's entry job into forestry. That, in truth, had been the case with professional forestry itself. Their fire mission obsessed early foresters: as much as anything, fire control defined the need both for public lands and for a force to protect them. With the help

of some well-placed fires and political timing, the Forest Service committed
to all-out fire suppression, and in 1960, when the Wallace District celebrated
the fiftieth anniversary of the Great Fires, the Service had no good reason to
look back except to measure its remarkable advances. Controlled burning had
virtually disappeared, save in logging slash. Fires that started from lightning,
arson, or accident were attacked relentlessly. Burned area plummeted. That
summer, though, Doc had to skip fire season to attend summer camp for his
forestry degree. He married, began a family, and returned to fire during his last
summer at Colorado State University, this time working for the Bureau of Land
Management as a firefighter and ranger. A year later he got on permanently
with the Forest Service on the San Juan National Forest in Colorado.

There is, he explains, no going back. His vigor will not spring back to his
bones; his career will not restart; the old 10 A.M. Policy that stipulated control
of every fire by ten o'clock the morning following its report will not revive, nor
should it. A more complex policy was needed then, as a more complex policy
now exists. The problem is that we cannot return to 1960 or to 1910 and replay
the past. The past endures in America's awkward policies and fire programs.
The past has shaped our understanding and political response to fire as fully
as it has sculpted the fire-disturbed forests of the Coconino. By the late 1960s,
it had become apparent that fire's abolition could be as powerful ecologically
as fire's kindling. The western pineries were disturbed as fully by the abrupt
disappearance of fire as the North Woods, cracked open by railroads in the
1870s, were by the reckless insertion of fire.

Like most foresters, trained to seek a science-based solution, he ignored
the past as irrelevant. Know the right principles, apply them, and the problem
will go away. But he came to realize, as the profession did, that history mattered.
The land looked the way it did not only because of what had happened but
because of the sequence of those happenings. Take fire, he tells the group. It
wasn't easy to change regimes. It began with overgrazing. In northern Arizona,
that commenced in the 1870s as sheep, driven by drought from California
and drawn by advancing rail lines, moved onto the plateaus and cropped off
the grasses that had sustained fire. Then came the simultaneous removal
of the American Indian, a prolific source of burning and a shaper of fire's

regimes. Well afterward, the land went into permanent reserves as public domain, principally as national forests that became committed, even fanatical, about abolishing fire from any and all sources. The three blows knocked out fire. By the late 1930s, burning was at impossibly low acreage. Then the rebound began, quickening during the late 1970s, until by the 1990s the grim scenes on the flanks of the San Francisco Peaks were typical of those throughout the once fire-frequented forests of the West. A century earlier the Norwegian naturalist Karl Lumholtz had commented on a profound paradox, that although "the Indians" of the Southwest and northern Mexico burned continuously, above the blackened ground the forests "stand green." The burning made them, Lumholtz believed, "indestructible" by fire. Now the inverse was true.

Doc lets the lessons sink in. He walks a little here and there among the dead sticks and fresh needles, probing with his cane, allowing the group to fan out and touch the crisp, fire-seared saplings, fondle the flaking bark of the deep-cooked, once old-growth yellow pine, kick gently at the duff. It's going to be harder than we thought, he says at last. What Chimney Springs demonstrated was that no easy reversal is possible. We can't put fire back in the woods as simply as we took it out. A lot of things happened to make this mess. A lot of things will have to happen to start correcting it.

And it's not just nature that's involved, he reminds them. You've got to deal with everything else in nature and people. There are invasive weeds and campers and bird-watchers and houses out in the woods now. There are people, a bunch of them, and what they think about all this—that will determine what might happen.

On the Gus Pearson Natural Area, part of the Fort Valley Experimental Forest, Doc opens a tall gate and strides along a path that began its existence as a fireline for a controlled burn. The gate is the single break in a tough, high fence designed to shield the site from ungulates, from deer, elk, or an occasional cow if any strayed that far. On one side, the plot borders a cluster of now-historic buildings, the residences and office of what was the first experimental forest established by the Forest Service. That happened in 1908. The other side of the plot fronts a woody jungle of pine, dense enough to hide a Mayan ruin, deep enough to smother a swamp. This is what the protected forest looks like ninety years later.

Doc lets the group walk a ways before stopping. You are in the middle of the prototype for the next generation of fire-restoration experiments, he informs them. The woods are in too big a shambles to accept fire as they are. Silviculture won't solve the problem, he notes—a tough admission for a career forester. The truth is, logging was part of what shattered the old woods by taking the big trees and allowing the brush and reproduction to reclaim felled sites and then arguing for the protection programs that stopped fires from flushing them clean. But fire alone won't solve it either, he reminds them. What you see here is a kind of synthesis, what we hope is the best of both. We've thinned out the small stuff, left the big boys, and started burning. This is what we think it was like a century ago. It's what a lot of us think all the woods around here ought to look like in the not-too-distant future.

Like many of his generation, Doc got into forestry to mingle with trees, wildlife, mountains. And like most of his cohort, he found himself dealing mainly with people instead. That admission dawned on forestry and the Forest Service with painful slowness. They tended to equate people with politics and politics with corrupt thinking and bad practices. But about the time Doc's career ladder carried him into the administrative rungs, the people decided they didn't like what the Forest Service was doing, and national politics, bolstered by litigation, forced the rangers to sit down and deal with lots and lots of people. The early critics wanted mostly more wilderness, more protection for endangered species, more recreation, less logging and ranching. Many didn't like the way the Forest Service handled fire either. Gone was the naive belief that rangering ought to be about engineering—experts making technical decisions and then applying them in the woods. Too many groups wanted too many things, and every one, it seemed, had a congressman or at least a lawyer at its elbow.

So Doc learned to deal with "the public." As a district ranger on the Shoshone National Forest in Wyoming, on the Wasatch in Utah, and on the Toiyabe on the east side of the Sierra Nevada, he increasingly found himself spending more time with people than with wildlife. Decisions flowed out of conference rooms, not the forest. For a while, fire protection seemed different:

an emergency response to a crisis. But suppression was clearly incompetent by itself. Some variety of prescribed fire also was needed, and that meant more negotiations with more people. It wasn't enough to keep people out and put fire in. It was a tough nut to crack, though, to decide what kind of role people should have.

He could use the demo plot as an example but leaves it to the group to work out the political geography. There it sits, halfway between untouched woods and a housing development. All three scenes are the outcome of choices people have made. Doc begins pointing out the features within the fenced plot—the stakes, the collecting dishes, the transects, the instruments, all the apparatus of modern research, for a nominally science-based program is the only kind Congress might fund. He explains the outcomes: greater biodiversity of ground plants and insects, more robust big trees, more open vistas. Then he comes to the gist. The inspiration for the prototype did not derive from ecological first principles or natural models or silvicultural experience. It came from history. Restorers saw the past and decided it worked.

Specifically, relying on their ability to date the living trees and various stumps (which rot slowly in this climate), researchers from Northern Arizona University unraveled the structure of the forest as it existed around 1870, before the onslaught of sheep, railroads, loggers, and the rest. The scene had been one of clumps of old-growth pine amid a veritable steppe of grasses and forbs. Perhaps twenty trees grew on an acre. With that as a target, crews "restored" the scene as much as possible, or at least its structure. They removed the small trees, most of which had generated in 1919; left the big ones; and spared an assortment of largish trees to make up for those that had existed in 1870 but had since expired. Then they instituted a burning regime, a roughly five-year cycle. The high fence prevented elk, in particular, from cropping off the lush ground cover and people from meddling with the experiments. Invasive weeds sprouted, then gradually succumbed to an invigorated flush of native plants.

Doc wanders over to a patch of untreated forest, left as a control. The group strains to see five feet through the woody thicket. The ground holds little save dense pine saplings, needles, and dead branches. Where the treated plot hosts twenty trees, the untreated bristles with nearly two thousand. It is, by any measure, a sick woods. Moving farther in, Doc comes to a residual yellow pine, pushing up through its root crown of needles, encased in a latticework of thick-barked saplings. Down it, like a crack of doom, rips an old lightning scar. The message is there for all to read: the decision about the future is not wholly ours to make. The fires will come. Which kind of forest do we want those flames to hit? What kind of fire do we choose?

That question—and the fact that Doc and others would ask it—comes from the experience of his own career. The public must decide what it wants to do on the public lands. There is no way to remove people from the scene. Among the biocentrics, however, a certain resentment simmers over the fact that the fire scene has become contaminated by people at all—by their camp-sites, their summer homes, their urban perceptions, their politics. People, they feel, have caused the problem by suppressing a natural force. People should correct it by staying out of the way and letting nature heal itself. Or if prescribed fire is necessary, people can assist by not whining about occasional escapes, smokey skies, and fleetingly ugly scenery. People get in the way. People make mistakes. People are the problem.

But it is trickier than that. The fact is, people have been on the land since the forest began taking its modern form after the last glaciation. Over the long millennia since then, the period of the late nineteenth century, when the forest was wiped clean of people, is the anomaly. A good fraction, if not most, of the missing fires of the last hundred years had been set previously through delibera-tion or littering by those humans; the flames had disappeared along with their tenders. That people recently are reappearing is itself a weird restoration to a norm. Whether people might return is not a serious issue: they will reenter the forest. They are already doing so, in ways both smart and dumb.

Doc is not so sure what role people had historically. In a sense, it doesn't matter because in a place like this fire would be abundant without people. He prefers to talk about the historic fire regime as "natural" and to sidestep the metaphysics, as he would a cow pie in a pasture. What makes the 1870 date so ingenious is that a determination of such particulars doesn't matter. Whatever swirl of forces converged to make a landscape of grassy parks and old growth, the biota was better off then than it has become. Besides, an appeal to history scotches the objection that this is only silviculture and logging in disguise, that restoration is really driven by commercial calculations. Instead, the scheme targets a time before rails connected sheep and timber with markets. Someone asks the obvious, Isn't the date arbitrary? Of course, Doc snaps. But any date would be, even one we set in the future as a goal. It just seems to me, he says slowly, through a sly smile, that we have to do something.

He leaves it to the group to survey the panorama, the buildings, the demo plot, the jungled woods. Whatever happens in the future will be the outcome of people making some decision or another, just as it has been in the past. His whole career has taught him that. Whether or not people are the problem, they have to be part of the solution, or nothing will happen. He looks from the prepared plot to the untampered one and back again. This is good enough for him. People will enter the woods. At the moment, he is himself leading a group of them.

This will be their last stop, Doc assures them, as he slides out of the van. He walks along a dirt road until he reaches an unfenced border. To one flank stands a swarm of burned tree trunks, black as a raven's wing, a flame-fried forest. To the other, a less-congested woods still redolent with a needled canopy.

A high-intensity wildfire blasted through a few years earlier. But before that conflagration arrived, some portions of the forest had been partially thinned and then prescriptively burned. The wildfire had ravaged the untreated forests, which lack only explosive craters to qualify as a scene from the Western Front. The treated area, though, had removed the source of the fire's fury. Like a hurricane pushing inland, the conflagration had lost its punch, had burned over whatever carpets of needles had been cast since the controlled burn, had torched a tree or assorted clump, but had not the means to surge through the crowns. The lightly slashed-and-burned plot had survived, the untreated had not. An aerial photograph of the contrasting plots made the front page of the *New York Times* as the 2000 fire season had raged on and the contours of a National Fire Plan began to form, in part around the proposition that some regimen of deliberate cutting and firing might be essential to certain of the West's public forests.

That, in truth, is where the architects of the demonstration plots would like to go. Doc explains that the program has a large-scale project under way at Mount Trumbull in the Arizona Strip. He doesn't explain that this is a place so remote as to be almost extraterrestrial—the Area 51 of fire management. The isolation means they can work through their trials without critics inhaling their smoke or breathing down their necks. The results will take years to sort through. They hope, more briskly, to connect to the intermix problem and are doing so around Flagstaff. With or without any controlled burning, a dose of woods thinning to something like historic standards will go a long way toward calming that volatile scene and to accustoming the public to the concept that some landscaping can help check wildfire. The National Fire Plan has proposed even more, perhaps a national campaign of industrial-strength woody weeding. But the devil, as always, will be in the details.

The photo and its instant-obvious lessons are, in fact, worryingly simple. There is no single protocol to protect every forest. Nor will such treatments eliminate fire, only change its regime to something like what the forests had known earlier and probably then only because people will reclaim the torch and do most of the burning. The photo makes good propaganda, the kind of political theater that pries loose money in an election year. The truth is that some kind of intervention is only the beginning, not the end, of restoration. Doc accepts that term, *restoration;* nuances of language and philosophy that would question its use rather than terms such as *recovery* or *regeneration* he brushes off as he would deer flies. He wants to see what actually exists on the ground.

So he doesn't lecture the group about having to choose one plot or the other. He isn't an academic: he became "Doc" as a navy moniker. He did acquire a master's degree in forestry, though, after he had retired from the Forest Service and taken up duties as a liaison with the Ecological Restoration Institute. He doesn't need to hold forth because his audience believes him. They listen raptly to his sometimes vernacular musings. He can speak quietly because he has a convincing authority behind him, the legitimacy of experience. He has lived through it all. The history lesson he conveys has less to do with this particular plot or that, this spared forest or that incinerated patch, a standard set in 1870 or 1780, than with the simple tenor of his presence, which is the complex sum of his own past. He ceases to talk and lets the group look for themselves. He doesn't end with a codified list of "history's lessons."

He doesn't need to, and perhaps he knows that history doesn't work that way. He was a better incident commander and a better ranger at the end of his career than at the beginning because he just knew more and could make better decisions, not because he had racked up a list of Principles and Lessons Learned from History. History meant experience; history's lessons meant a better-informed judgment. History never seemed to present the same scene twice, so why should knowledge gleaned in one moment apply unvarnished to another?

So it is with nature. Every site is different. Sure, there are general principles, but they have value only if they are shaved and whittled to fit the particulars, which takes appraisal and tact. And if people are involved, as they must be, all technocratic bets are off. All this takes judgment, which can happen only when knowledge and experience have passed through the mind and emerged as something resembling wisdom. That, in fact, is what the Pearson plots convey and the *Times* photo should confirm: the history that must guide our choices will not be encoded in "lessons," "expert systems," "simulation algorithms," "prescriptions," or other technocratic blather about learning from the past. That past will speak only through particular places and in particular thoughts. Above all, if the past is to hand on its experience, it will speak through a human voice.

That should do for now, Doc says. The tour has ended. They pile into the van. There are many such tours to make and thousands of decisions about what to do with this patch of wildland or that stretch of woods. There are other researchers, other sites as vital to the future rehabilitation of open fire, other prescriptions and tangled histories, other guides and oracles and prophets in one wilderness or another. If they are wise, or at least shrewd, however, the groups that listen will understand that the lessons of history they have learned reside less in the fire-scarred slab Doc held than in the hands that gripped the slab and in the mind that guided those hands. Somewhere, the choices to be made will have to merge into a voice. If woods-goers are lucky, the voice they hear will sound a lot like Doc Smith's.

back
at
the
cache

smokechasing

the
search
for
a
usable
place

The wind drives the rain in sheets, rippling like dense drapery. Raindrops the size of pine cones splash off the pavement. We sit in the fire cache, crouching on milk drums and rounds of ponderosa pine, our fire packs at our knees, roughly grouped around a fuel-oil heater. Lightning flashes, not far distant. Another bolt cracks in a long rumble that probes across the Transept like the tendrils of a vine. It's difficult to believe that fires can start amid such conditions, much less thrive and inspire a crew to hike in and extinguish them. But we know they do. Not here, of course, not just outside the cache's great gated doors, but beyond the pale of the unwashed windows, beyond our huddled selves, a fire that will send us, all of us, the North Rim Longshots, dashing to an engine with a holler and a hope. That's why we now sit and wait for a smoke report from Recon 1, our aerial observer. We wait; we listen; we, those of us who have been on the Rim for a few seasons, think on what we have learned and why we are here.

The short answer is, we are here because of fire. No fires, no fire crews. Our lives arrange themselves around wildfire much as we now cluster around the virtual fire of the heater. But none of us is solely interested in fire alone. Smokechasing—the practice of finding and attacking wildland fires—is not just an adventure; it's a life. We relish the place. We aren't swatting out fires in a city's vacant lots or in a fallow field: we live on the Rim of the Grand Canyon. Sure, the place would be much less interesting without fire. Without fire, we might have to live as fee collectors or ranger cops or fern-feeling naturalists. But the fires are there for anyone who chooses to pursue them. Fire season is a time in our lives, for which the Rim is a geographic expression. A fire on the Rim is what makes the site habitable and our time there meaningful, what makes this life the best of all lives in the best of all places.

For some of us, smokechasing can trek beyond the Rim and become a model for life: the smokechaser as scholar. It's not for everyone, nor is it the only way to learn about a larger world. Scholarship is more deliberate and less physically taxing, although I know more people who have broken their health with a pencil than with a pulaski, who can trace more ailments to a flickering computer screen than to a flaming snag. No shovel-wielding Longshot ever suffered from carpal tunnel syndrome. No crew flush with fires ever sank into clinical depression.

For someone attracted to environmental history, the transfiguration of smokechasing into scholarship comes seamlessly. History has been described famously as the search for a usable past. Environmental history adds to that: the search for a usable place. That's a smoke report I can rally to. I keep a fire pack for research travels on the shelf, loaded, ready for dispatch. I have put fire at the center of humanity's existence, much as it was for the Longshots on the Rim. By themselves, the fires are not enough, but with a suitable place to site them, they can make a life. They have made mine.

Outside, another burst of hard rain and a blast of lightning whose flash and crack are one. The rookies in the cache think back on fire school, trying to recall the sequenced lessons, the Ten Standard Firefighting Orders, the lengthening lists of Situations That Shout "Watch Out." I'd like to say that everything I needed to know I learned in fire school. In fact, two or three days of classroom instruction mean little. They mean we get a certifying document, a red card (the Park Service's were pink, I never learned why), which allows us to go on a fireline. The lessons, however, are a bureaucratic scam unless followed by labor in the field. The only real learning comes by doing. A parody of the Paul Simon song "Kodachrome" (revealingly renamed "Overtime") would make the rounds annually: "When I think back on all the crap I learned in fire school, it's a wonder I can work at all."

The point is to get into the field. A sign hangs over the cache, a warning to those who enter: "If you don't get outta here, you don't get outta here." Begin the morning by hanging around the cache, and you'll hang out all day. There is always a canteen that needs refilling or another memo to file. Start in the cache, stay in the cache. The value of a fire, a real fire, is that it forces you to get out. No Longshot would want it otherwise because it is always better to be on a fire, any fire, than to potter around the cache sharpening axes, linseeding shovel handles, or washing the three-quarter-ton Chevy. Better to be called in to help mop up a fire. You get to smell smoke, do some real work, and maybe claim some overtime. Still, mopping up is never as good as having your own fire, however wretched or magnificent, be it smoldering duff or a fully flaming ponderosa. Every rookie should be rabid for the adventure of tracking down that smoke, for the immortality of naming it, for the thrill of calling in and directing others (including the old hands) to do the grunt labor of cutting line and mopping up. The trick is to get a big fire, but not too big—not so big that outsiders arrive to take it over and absorb it into the bureaucracy of big burns and relegate the smokechaser to smokestomper. When that happens, it's time to return to the cache, replenish your pack, and wait for another call.

Our usual dispatch procedure is to send out a veteran and a rookie together because that's the only way the new Longshots can learn, because, very shortly, they must become vets themselves. A bad crew runs by seniority. Those with experience and standing claim all the fires, sop up most of the overtime, and leave the mop-up to the newcomers. Politics and bureaucracy take over, however packaged in pedagogical cant about mentoring. Seniority rules when there are more firefighters than fires, when the bust is not big enough to send everyone, when someone has to remain and tend the cache. Better, rather, to be stretched than flabby. Better to have a crew too small than too big.

There is ample precedent that the same holds for scholarship. The last great round of scholasticism, in the fourteenth and fifteenth centuries, also slopped over from France. Its essence was a glut of critics and a deficit of texts. The upshot was a deepening tar pit of glosses and glosses on glosses until the texts themselves dissolved amid the acidic sludge of rhetoric. Worse, the practice had the sanction of the establishment. To argue against the prevailing premises of the Schoolmen was to risk charges of heresy. The risk was there even if one did not attack the reigning dogma directly but simply ignored it. One could not speak against the protected categories, nor could one speak without reference to them.

What finally broke thought free was a determination to read the Book of Nature, which was really not a text at all and which therefore demanded a *novum organon*, a new logic, the methods of modern science. The authority of Nature challenged the authority of the Ancients. The antitext of Nature was an antidote, intractable and inexhaustible. The true Otherness of Nature defied rules of rhetoric. Nature proved unbounded; Nature frustrated intellectual bureaucracy; Nature blasted constructed categories with the violent indifference with which lightning strikes stone and snag equally. Moreover, science acquired additional power because it bonded with technology. It was tested not merely against logical consistency and ideological templates but against whether or not it worked.

The contemporary counterpart to scholasticism is postmodern cultural analysis, a guild that has too many apprentices for too few jobs. The one is a condition of the other. A profession that requires eight, nine, or ten years to complete a degree is training mop-up crews and cache workers, not smokechasers. Its apprentices don't make history; they study historiography. The solution is to get into the field. The way to remove scholarship from the Möbius strip of rhetorical scholasticism is to turn again to the natural world. A guild cannot control nature as it can texts, canons, or degree requirements. Whether environmental history can evolve a novum organon is doubtful, but it is ideally situated to strike off the shackles of a miserable solipsism and a stupid scholasticism. Nothing else comes close. But having originated in the Earth, the discipline ultimately must return to the Earth. The test of its rightness will be whether it can site a usable place.

That understanding is intuitive with the Longshots. We know that release is there—that nature sets fires, if only according to its own ineffable logic and whim; that we are not left to endless manuals, computer screens, drills, and memories. Our task is to seek those smokes out, which we do by sending a Cessna 172 aloft to dart through thunderheads, by staffing Kanabownits lookout after hot storms, by pausing as we drive to the lodge to scour the eastern horizon, the limned bulk of Walhalla Plateau, for smokes, by clambering up to the crow's nest of Tree Tower 1, atop a blasted white fir. "You can't wait for inspiration," Jack London once wrote. "You have to go after it with a club." We go after it with binoculars, a compass, and a shovel.

Two smokes are reported on Swamp Ridge: Bill and Sandy scamper to Engine 652, and, because it's an early report, Tommy Deshennie Begaye, one of the Navajo Southwest Forest Fire Fighters (SWFFFs), goes as well, while Barry and Fran and Henry Goldtooth race off in 653. Get those rookies out of the cache and onto fires. The bust is on.

The storm cell continues to boil out of the Canyon, its crashes rumbling from the west. In its wake, Recon 1 reports a possible smoke on The Dragon. The wisp of a groan rises from the Longshots. It's not that a fire on The Dragon is unexpected or unwanted. Acre for acre, the mesa, almost wholly cut off from the Rim, has more fire than anywhere else. This is prime habitat for a fire crew and a place known to them. The problem is that The Dragon's fire has become problematized.

The Dragon is an ideal setting for what the Park Service calls a *prescribed natural fire* (PNF), which is defined as a fire kindled by natural means that burns in a predesignated place under approved conditions known as a *prescription*. This is a dictionary not an operational definition. No one knows what it means in the field. What it means on the Rim is that we staff the fire but don't fight it. Or that we fight it and try to disguise that fact. Or that we ignore the smoke—let it blend into the greys of the Kaibab limestone—and congratulate ourselves for our creative restraint.

The route to the PNF is as overgrown with good intentions as the trail to The Dragon is with black locust. The PNF emerged as a device to let a natural process seek its own destiny yet still have a political bureaucracy maintain control. It allowed us to put fire back into a biota that craved it, yet to assign responsibility for the outcome to nature. We could be there, yet not be there. We could install natural fire as the pith and purpose of fire management, yet justify intervention where—for reasons of scale, politics, or past history—the PNF proved impossible and plain old prescribed burning would have to substitute. As the years rolled by, the National Park Service, inspired by the PNF chimera, burned up almost half of Yellowstone at a cost in excess of $130 million and, when PNFs became suspect, succeeded in burning a chunk of Los Alamos in the name of naturalness at a cost in excess of $661 million.

The PNF was a cultural creation, the invention of a society that had polarized the world, that said we either had to fight fires or let them burn and preferred that we do both at the same time. This was a society that had no working concept of a usable place. Yet what critics condemned as a contradiction and advocates defended as a compromise, the philosophically gifted might recognize as one of modernism's many paradoxes. The prescribed natural fire joined Russell's paradox, Gödel's proof, Bohr's principle of complementarity, and Heisenberg's principle of indeterminacy—all of which struggled to incorporate the observer into the observed system. The question whether such fires were natural or controlled was of a piece with the question whether electrons were particles or waves. (If physicists can cope with electrons, surely historians can cope with events that are both natural and cultural. If they can build bombs, we can write books.)

The philosophically challenged might note that cognates existed in popular culture as well. Over the winter of 1967–68, during which the National Park Service reformed the policies that led to the evolution of PNFs, *Star Trek* began its TV serialization. Both Starfleet and the Park Service accepted the same Prime Directive: noninterference in the evolution of other life forms. If somehow that order was violated, then it was acceptable to intervene to restore conditions to what had prevailed previously. This describes almost exactly the philosophy of wildland fire management at the time. The PNF was Starfleet's mission to an environmentalist Earth.

"Please, please, not another Recapitulation fire," Scott mumbles. Last summer lightning blasted an enormous snag on The Dragon, part of a far-flung bust. Scott and Fran flew to the helispot on the north flank, then hiked in, felled the snag, corralled the flames with a fireline, and hustled to another smoke, while a couple of South Rimmers replaced them for mop-up. Park officials, however, squirmed over the episode. They couldn't leave the fire, yet neither did they wish to engage it. So they secretly instructed the South Rimmers to repair the damage the Longshots had inflicted. This was a natural—a wild—place. There was no justification for displaying the human presence. We should remove the history of our unwise acts. So the South Rimmers flushed the stump, burned up evidence of cut logs, and rolled back the pine-needle berm that rimmed the fireline. Then they flew out. Two days later a large smoke blasted out of The Dragon. The fire had rekindled and overrun the restored fuel. Another squad of Longshots attacked it, felled more snags, cut more line, and fired out the interior. This reestablished control but left unresolved the question of our continued presence. So instead of spading every smoke into submission, the crew sat around for two days munching on rations, sleeping under the pines, and reading broken-spined paperbacks until the fire expired.

The Recapitulation fire had become large enough that it demanded a narrative report, not merely a coded form. But how do you write such a narrative? It is not apparent how we should be on The Dragon or how we should tell the story of our being there. If we exclude ourselves completely, then we have no story. If we only observe, then we end up with stories about ourselves observing, which means in practice stories of ourselves observing ourselves. Yet if we center the story wholly around ourselves, then we lose the special, unsettling power of natural fire and The Dragon. Instead, the narrative becomes an exercise in justifying theory and its expression as policy, which by definition cannot be wrong. Eventually the concept of the PNF imploded, no longer an accepted strategy of fire management and stricken from its working vocabulary. What should replace it is not clear. The 2000 fire season witnessed a wholesale breakdown: we couldn't control fires we started and couldn't suppress those nature kindled.

Such matters concern the smokechaser as scholar. The PNF flourished in the same years as the breakthrough books of environmental historiography. Precontact landscape and, in particular, wilderness proposed a sublime and brutal contrast with American society and suggested—maybe demanded—a new strategy for doing history. Certainly it stopped the usual practices: it was no longer acceptable simply to suppress nature from scene and story. But the polarity, however majestic, was too great to sustain, and after the initial shock it was not obvious what kind of narrative report historians should file. To remove ourselves—accept a let-burn strategy—is to yield to ecological history, a biotic chronicle with people rubbed out or stripped of moral agency. To continue to bumble around—pretending we are both there and not there—likely will spawn historiographic fiascos akin to the Upper Frijoles Canyon Number 2 fire that bolted out of Bandelier. The easiest, default solution is to center the story around ourselves.

That is what humanistic history would like. It could absorb what passes for environmental history within its prevailing operating system—the codings for race, gender, class, and ethnicity; the Windows of academic history—and then could apply the same strategy that Microsoft has used against its competitors: embrace, extend, extinguish. Environmental history as a distinctive project will be subsumed, sidelined, or squashed. It will be shrunk into body studies, surgically transgendered, strip-mined for environmental justice, dissolved into a misty sense of ethnic place. Because discourse resides within institutions, institutions will become the object of the discourse. Historians will study fire policy and park bureaucracy. The Dragon will exist as a problem, not a patch of land. All of this is scholarship, much of it good stuff, but such history will thrive at the cost of abolishing what made a fire on The Dragon special—that is, not us, but fire and The Dragon.

"Maybe," Scott mutters, "Recon 1 saw a waterdog." Recon 1 isn't sure. There are enough true fires popping up that it isn't worth committing a crew to The Dragon just yet. The issue of how we should "manage" fire there remains deeply, deliberately ambiguous. If such fires are all we have, we will dissolve from boredom and paralysis. The Longshot position is that if fire belongs on The Dragon, we ought to put it there. We ought to burn it ourselves. The story should be of us doing just that. But this would be too simple—would confuse a natural process with an unnatural act, would apply a pragmatic logic to what the reigning authorities obviously cherish as an unresolved and unresolvable *problématique*. Instead, debate will replace decision; power will shift to those skilled in the arts of discourse. Fortunately, there are plenty of fires elsewhere. Recon 1 spots one in The Basin.

The deluge pauses, catching its breath. The inexperienced Longshots nurture the belief that the storm is ending. The older are more wary. They know storms beat to their own logic and rhythms. All turn their thoughts to the mechanics of those in the field and rehearse the stages of smokechasing. Recon 1 has flown over Engine 652, parked on the Swamp Ridge Road, in a direct line to the smoke, about two miles distant, and reckoned a compass bearing of 153°.

The bearing by itself means nothing. What matters is its correction. You learn quickly—a single fire is sufficient—that Recon 1 flies on true magnetic bearings such that "north" on its compass points to the magnetic north pole. But the maps that guide smokechasers point to the north pole of rotation. On the North Rim, the difference between map and compass is 15° east, a distinction known as magnetic declination. The gap matters. Even over a short distance—a mile, say—a 15° misreading could put you on the wrong ridge. On the North Rim, where the terrain resembles a corrugated roof bent into a shallow dome, all heavily forested, to follow an uncorrected bearing is to get lost, and amid the rush of dispatching that happens annually. It will continue to happen so long as those observing and those responding have compasses set to different norths. Within that difference, crews at night can lose not only their fires but themselves.

The Longshots have many names for this disparity, most of them unprintable, but a good literary term for that gap is *irony*. It's not a word you hear often on the Rim because it's not a usable concept. It is not, ultimately, an instrumental word that belongs in a fire pack with flagging tape and rations or that can be hoisted on a shoulder like a chain saw. On the Rim, irony is little more than a bluster of wind because, once recognized, its defining gap can be sealed. In modernist scholarship, however, irony endures as a structural condition much as declination persists as a shaper of narrative. No one, it seems, wants to close that chasm.

From the beginning, irony has been the voice of modernism and has thrived on the great gaps of modern times—between expectation and reality, between the promise of progress and the atavistic horrors of the twentieth century, between a world of tradition and one of modernity, between nature and culture, between people and place, between the world we say we want and the world we actually make. But these are ancient lapses. What is distinctive about modernism is that it refuses to grapple with that gap, to close or bridge it. Instead, modernists typically have sought either to eliminate the chasm by razing one side completely or to hold the chasm permanently open. The former has inspired some of the worst absurdities and vast terrors of the past century.

The latter leads, limping and huffing, to the vapid voice of postmodernism. Postmodernism is irony playing tennis without the net.

Much of environmental history—like nearly everything else in the modernist panorama—is a literature framed by irony. The distance between culture and nature is ideally suited for ironic commentary. Think of the contrast between the wild and settled. Think of stories of contact and colonization where the distance between rhetoric and reality tends to be huge and the ditch between people and place can be as wide as the Atlantic Ocean. So pervasive is the ironic voice that it shapes the stories one might choose, selects the frames, polishes the phrases. Irony is the default setting of academic history. No one is chided for speaking ironically, but not to speak with an ironic edge is to risk charges of naïveté. Once a discussion has been framed ironically, it typically ends. Irony knows no back of beyond.

It makes good banter in a cache while the crew is seated around a heater, holding coffee cups or outfitting axes with leather sheaths. (I say this as someone who has made a career as a serial ironist.) But irony is worthless, even fatal, in the field. It belongs in the cache like safety posters, not on an engine. Those at a desk have the luxury of pondering whether one can ever truly "find" a smoke or whether, over the long run, one can "suppress" flame. Those in the field huffing up another damn ridge, soaked from brushing past soggy firs, see it differently. They prefer to keep irony as a kind of mental whistling—what Recon 1 describes as a small hill may well seem to a groundpounder like a miniature Matterhorn. But no Longshot would accept seriously irony's vision as a bearing to take into a smoke. No smokechaser would be rewarded, as a contemporary scholar might, by showing the futility of smokechasing. No fire, no firefighters.

Irony requires, ultimately, an *un*usable place because if people reconcile themselves with the land around them, the gap closes and irony becomes irrelevant. They set their compasses to a common standard (it doesn't matter which, only that it be the same). Like modernism, irony has shown itself superb at dismantling old structures and has proved wholly incapable of building anything of its own, for that would demand that it close the gap that makes its vision possible. Yet the modernist project requires a cultural breach that now exceeds its gasp. Irony has enjoyed a monopoly, and like all established monopolies it has become predictable, oppressive, boring. In firefighting terms, irony is no longer even a sculpted snag but a smoking root hole. A smokechaser compassing through the woods at night with a chain saw on his shoulder doesn't have the luxury of contemplating the unresolvable *différence* (or *différance*) between privileged bearings. You can get very lost or very hurt. Maintaining that gap becomes either an indulgent game or a cruel exercise in power. It's time for a postironic culture.

In fact, the public is out there, oblivious to the philobabble of postmodern criticism, its sleeves rolled up, doing things in the dirt to make a usable place in which to live. Historians can point out that, granted irony's implacable status, such efforts are doomed, or we can join them and write histories that are the intellectual equivalent of ecological regeneration. To do that we will have to remove irony from its privileged pulpit and have it join the choir, take it out of our mis-set compasses and put it on the tool racks with the pulaskis and korticks. And to do that we have to scrap the protected categories of thinking that make irony inevitable. This is why *ecological restoration,* for example, is a regrettable phrase. Yes, its use is pervasive and its pedigree honorable; sure, it carries real cultural clout—that we are making good what we have wrecked, that we've seen the past and it works. Yet *restoration* locks us, again, into an ironic narrative. We can never truly restore a place. What matters is making where we live habitable. Better to speak of ecological recovery or ecological regeneration. The point is the future, not the past.

Another smoke is sighted along Swamp Ridge near the Tipover tank. Randy and Jan and John Paul roar off in a borrowed ranger pickup parked under the big white fir by the hose rack, hastily outfitted with extra tools and rations and sleeping bags wrapped in plastic garbage bags. The rear tires spin as they grab for the edge of the pavement. So, in truth, the world rushes over and scatters our categories like so much loose gravel. Irony belongs with those loose pebbles by the road and the occasional cracked windshield.

The lightning seems more remote, the rain more feeble. Recon 1 has veered north to escape a ferocious thunderhead over Saddle Mountain and spots a smoke to the northwest, near VT Park on the North Kaibab National Forest.

He reports it to the Forest Service dispatcher in Williams, who requests that the park respond. An off-park fire: always an adventure. Any fire outside the park is better than no fire within. Our vehicles, however, are getting scarcer. One Longshot, Larry, and two SWFFFs wheel out in a stake truck borrowed from maintenance, more often used to haul garbage and asphalt. They leave with Larry defiantly holding up a pocket notebook to record his anticipated overtime.

Sometimes, though, we travel farther. We go to the Strip, sometimes to regional parks such as Zion and Mesa Verde, occasionally to California. Going off-park is another way to test our skills in new settings. A Longshot has to demonstrate that what he or she knows matters elsewhere. So, too, a smokechaser-scholar has to travel out of academic departments, beyond the borders of defined disciplines. Or as William James, waving a different kind of notebook, put it, you have to ask what in experiential terms the "cash value" of our ideas really is.

This can get sticky for disciplines such as history, grounded in the humanities, that in the final tally create and analyze moral universes, that address questions of who we are and how we should behave. Surely, nature is amoral and stands outside the purview of humanistic scholarship. What value is the humanities to natural history or natural history to the humanities? A lot. The humanities have value, first, because creatures besides humans make choices about their behavior and hence exercise judgment—which is to say, even non-human nature is not wholly removed from moral realms. Because, second, nature prompts people into moral agency. It forces us to choose—catalyzing and compelling decisions and acts—and thus becomes a moral presence. Its human-shaped landscapes record those choices.

Because, finally, ecology is a historical science like geology that cannot explain why landscapes look as they do except by appealing at some level to history, and because it cannot ignore what is often the keystone species in the evolution of landscapes and still hope to offer more than fanciful tales of how and why it all happened. An environment without people is even more abstract and meaningless than the ideal frictionless surface beloved of physicists. A place inhabited by humans ought to be the norm for analysis and a place emptied of people should be considered an outlier, an exception, not the ontological basis for understanding. One might as well shut off rains or sunlight, either of which also would simplify calculations and flowcharts. But to put people into those places means one must understand how and why people act as they do. That the natural sciences cannot do.

Fire demonstrates this nicely. We are born genetically equipped with the ability to manipulate fire, but we do not come programmed knowing what to do or how or why. And because our capacity over fire is a species monopoly, our fire behavior serves as a unique index of our own ecological and moral agency, a measure for which no neutral position is possible, for fire can be as ecologically powerful removed as applied. If that isn't a subject fit for environmental history, nothing is. If environmental history can't speak to the needs of fire management, then it is truly and pejoratively academic. In fact, it can, though it largely has chosen not to.

The last time it spoke to those needs for me was in 1998 when I went off-park to Ghana. The two sponsoring agencies—the Forest Service and the International Tropical Timber Organization—didn't want a historian. They wanted a sociologist. They wanted some technocratic brand of scholarship that could address the fact that people were behind Ghana's fires. They wanted "methodologies" and "manuals" and the authority of science—its political if not epistemological authority—even if they had to seek for it in such troubled disciplines as sociology, which in any event had no one up to the task. So they settled on me.

Actually, I could do much of what they needed. I could tell them how America had evolved from a rural, fire-saturated landscape to an urban, fire-starved one; how Australia had sought to reconcile rural traditions of burning off with a search for national identity; how Europe seemed congenitally unable to conceive of fire as anything but a social creation. I could tell them why, based on historical evidence, fuelbreaks worked and failed. I could recommend how to segregate science from management, how the political ecology of fire worked, how institutional inventions such as forest reserves have survived the withdrawal of the colonial powers that installed them. I could tell them why they didn't need a formal questionnaire, only suitable questions.

As a scholar grounded in the humanities, I could read with skepticism the text of a Netherlands-sponsored proposal for a ten-year project in bushfire control. The Dutch wished to erect a system of cultivated "green firebreaks" around the threatened forest reserves. The literary critic warned about the choice of words and imagery. Would the scheme have the same punch if those barriers were called "cassava corridors" or "plantain patches"? Would it have the same political panache if it were not premised on the vision of the Sahara marching to the sea? The philosopher noted the many political values embedded in the coded clichés of three-year plans and sustainable agriculture and masked in ambiguous phrases about holding village lands in trust. The historian identified the persistence of European ideals about landscape, of beliefs about the relationship of trees and climate, of concepts such as "desertification" that have survived, tough as spores, in the face of hostile evidence for millennia. The cultural critic observed that the Dutch were treating the shallowing forests of Ghana as they have their own shallow seas, that the firebreak barrier was a kind of seawall, that the intricate network of green fuelbreaks was a system of dikes by which one could drain fire from the landscape. All this the Ghanaians needed to hear.

But ultimately the humanities need to hear Ghana even more. It was only by default that the humanities were there at all. The recent humanities seem to see nothing except people of specific categories and texts of unspecified abstraction. But semiotics is not consuming *Chromolaena* and mahogany; combustion is. An ironic voice cannot say whether, on a February afternoon, flames in three-year-old fallow will hurl embers over a forty-meter-wide fuelbreak. Gendering cocoyams and maize does not determine when forest guards should early-burn along the slashed boundary of the Pamu-Berekum Forest Reserve. Wildfire does not care whether the colonial era privileged indigenous forest or exotic teak plantations. Drought comes regardless of ethnic choices between shrubs and woods. Play with fire—real fire—and you can get really burned. The humanities and much of environmental history that derives from such scholarship stand apparently helpless before those flames. What burns away finally is history's own presumption that the world needs it more than it needs the world.

Every fire crew knows that dispatches end with returns, that you aren't paid until you submit your report. The paperwork counts. The written word is what survives and what the bureaucracy believes. Yet a report by itself is worthless or, if submitted without that informing smoke, fraudulent. The report—and all the appended memos attached to it as it migrates up the institutional food chain—might by itself be good enough for some endeavors. It's not good enough for smokechasers because they know that what matters is the flame behind the fortune, that the money is only an excuse for the experience. Those wildfires are not on the Rim because of us. We're there because of the fires.

The downpour abruptly turns to hail and lashes across the metal roofing of the cache like a sandstorm while rekindled lightning causes the lights to blink. Ralph the rookie worries that the fire bust has ended without him. The veteran Uncle Jimmy mutters that there are always fires if you know where to look and keep the faith. You just don't find them on the shelves next to the National Park Service administrative handbooks or amid the chatter of the Grand Lodge saloon. Ralph replies hopefully that, if nothing else, they might be summoned to bring in supplies and help mop up one of the existing burns. It could be worse. That's when Recon 1 reports a fire south of the Point Sublime Road, near the Rim.

The smoke is puffing up near an aspen grove, not a good omen. But Recon 1 believes it actually is coming from the base of a big yellow pine. Well, any fire is better than no fire. They shoulder their packs. Besides, Sublime is hours away. The storm could break, the wind rise, embers spill over the rim into ravenous chaparral. That much nature would determine. What the Longshots make of it is up to them. Ralph has already decided to name it the Genesis fire, in the hopes that it might become the start of something special. Uncle Jimmy rechecks their hand tools.

It's what they live for. Ralph nods and shows his newly acquired fire-school lore by declaring that they had better hurry, that the point is to get to the fire before it gets too big. But Uncle Jimmy, wise in the ways of the Rim, shouts his disagreement over the hail. The point, he insists, as he scrambles out the doors, is to get there before the fire goes out.

almost
lost

It was the end of a long night of a long day of a very long research trip, and I was not at all pleased to be locked inside the men's restroom of the KOA campground outside Austin. It was almost midnight; the campground was largely deserted, and the likelihood was that I would remain in that cinder-block cell until some bleary-eyed, unshaven camper rattled the door at dawn.

I was there—Why? I asked myself. Because I had a cooperative agreement with the Forest Service to write a history of fire, and we had acquired an ancient International pickup with a camper shell retrofitted into an office and a secondhand, single-axle Timberline travel trailer to carry us around the United States while we hopped from archive to archive. Because Sonja, my wife, had flown back to Phoenix for her brother's wedding. Because someone had to haul that rig from Florida to Arizona. Because, weary at the prospect of trekking across West Texas alone and weakened by two years' absence from any kind of serious talk, I had decided to lay over for a day at Austin and visit old friends and professors at the University of Texas. Because things like this just happen.

The evening had passed quickly. We told some stories, drank some beer, reminisced. No, I confessed, I had no academic prospects, not even the rumor of an interview. No, the cooperative agreement did not pay me a salary, much less anything like benefits, only per diem when I was on the road. Yes, I still worked summers on the North Rim, foreman of the Longshots, and would continue so long as knees and heart held out. Yeah, I was doing what I wanted, what I liked, what I chose. They told me about the latest departmental gossip—who was in, who was out, who had flunked orals and who had a job offer, who had written a few more chapters. I told them the saga of the Miss fire on Tiyo; of the whirling ponderosa that nearly squashed Robert John and his banana hat; of the hilarious, compromised trek to the Atoko Two fire; and of how I rescued thirteen boxes of mimeographed minutes and old reports from incineration at the fumbling hands of the Washington, D.C., offices. They explained, through a creeping fog of beer, the conventional wisdom of a campus-bound grad student: it's not what you know but who you know that matters. Crashing through the woods with a pulaski and a compass does not constitute academic networking. Searching out smoking snags on the Kaibab Plateau does not make a dazzling c.v. I was selling the family cow for a handful of magical beans. I had to get back on the inside. Yes and no, I shrugged. It's what I choose.

I drove back to the campground. I could see a banner of stars overhead as I trudged to the men's room and punched in the code to open the locked door. Nothing happened. I should have stopped right there. Again I punched in the code; again, nothing. I pushed the buttons once more, very carefully. This time the door opened. But it refused to budge on my way out. On the inside, there was no numeric pad to fiddle with—only a handle to pull open a door that was obviously jammed with a broken lock. I was stuck.

I stared at the door. A mixture of outrage and despair welled up, then dissipated, crowded out by the too-familiar sensation of hiking into a smoke at night and getting lost. I had a knife in my pocket and a roomful of steel hardware, though, so I went to work. I unscrewed the door handle off one of the stalls and scraped the flat end over the concrete floor until it had an edge. That gave me a crude chisel. I undid the faucets from one of the sinks and soon pried out enough metal to make a mallet. I wrenched the handle off the entry door—a heavy, solid-core hunk of glued wood chips. I then hacked away around the lock assembly until I could pull out the dead bolt and the locking mechanism. After maybe an hour, the door swung free. Next I disassembled the damned numeric keypad. I laid the entire apparatus out on the floor and walked to my trailer.

The next morning, bleary-eyed and unshaven, I made my way to the men's room. A small crowd had gathered around the entrance. They were staring down at the dismantled lock and chisel and lead-pipe mallet, and one of them, apparently the manager, was kneeling beside the display. As I pushed my way in, I heard him mutter, Yeah, but it looks like it was taken apart from the *inside*.

I said nothing, expelled the last of the beer, and drove, day and night, back to Arizona, my head full of fires and books.

an
exchange
for
all
things?

All things, Heracleitus declared, are an exchange for fire, and fire for all things. It was the kind of obscure utterance that earned him the sobriquet "the Dark." Yet the pronouncement suggests that fire might somehow serve as a universal currency, a gold standard for natural history, not only for how nature works but for how we might understand those workings. Nothing of the sort has occurred. Fire is seemingly everywhere and nowhere. There is no academic department of fire, no discipline devoted to fire studies, no understanding of fire itself as a universal referent, as a common process with which other subjects might be exchanged.

Why? Why should fire, which has been on Earth for several hundred million years, not command an intellectual domain of its own? Why should sentient humans, who after all possess a species monopoly over the manipulation of fire, ignore something to which their myths relentlessly attribute their special status? How can so much and so little be known about fire? Why should so many fire inquiries end in scholasticism, a welter of theses and countertheses and glosses without clear resolutions? Why is the once-central fire no longer central?

One explanation is that fire is a chemical reaction, a biotic catalyst that acts on whatever exists around it. As in nature, so in scholarship: fire cannot exist alone. We know it not in itself but by its relationships. We know it by how it functions within grasslands, forests, fields, pastures, soils, watersheds, airsheds, and the like. So also we know it through the scholarship that informs our understanding of these phenomena. We know fire within the context of ecology, botany, forestry, geography, zoology, meteorology, chemistry, anthropology, even (my field, say it softly) history. The arguments about how fire behaves in the world is, in this sense, a quarrel among disciplines over a subject that they all share but none owns, that acquires meaning only from the context of one or another discipline. What a forester accepts as experimental evidence a physicist may scoff at. What an anthropologist accepts as convincing an ecologist may dismiss as anecdotal. Certainly the literature on fire breaks down along these lines.

But another explanation is also possible: that fire research reflects historical geography. What is studied, how, and by whom result from peculiarities of history played out over particular places. Who studies fire? Not all people do. Where do they study it? Their residence and the fires may be continents apart. To what purposes do they study? Those who use fire are not always (or often) those who study and judge it. In their simplest formulation, fire studies require both modern scholarship (which means mostly science) and abundant fire on the land.

Yet even a casual glance at the distribution of fire on Earth shows that only some countries possess both. Most of the world exhibits what satellites positioned over Europe and Africa reveal, that the science is in one place and the fires in another. One landscape shines with a dense constellation of electric lights, the other with a tapestry of flame. That handful of places with both science and fire are the product of an Enlightenment Europe that colonized widely amid fire-prone landscapes and left a residue of public (or Crown) lands behind and a bevy of institutions to administer them, largely under the control of foresters. Here reside the primary centers for fire research—the United States, Canada, Australia, Russia, and, less securely, South Africa. Where fire is studied in the rest of the world, the research works through institutions dispatched from these centers or from Europe in a kind of shadow colonization by Western science.

Thus, the grand curiosity of fire scholarship. The fires appeared in one place, study of them in another; those who exploited fire failed, overall, to analyze it, while those who wished fire to disappear became the guardians of its scholarship. Whereas most burning remained within agriculture, conducted by swiddeners and graziers, understanding of it fell almost exclusively under the dominion of foresters, who detested slash-and-burn farming, distrusted fire-trailing pastoralists, and disliked open flame of any kind, for whom fire exclusion remained an ideal, however quixotic. Thus, the study of fire fell to a group committed to abolishing it. It was as though anthropologists studied primitive tribes in order to remove them, or zoologists studied penguins only in order to eradicate them. This had, needless to say, considerable consequences for what would be studied and how it would be interpreted.

Forestry made an odd niche for fire. It was narrow in its disciplinary focus, yet global in its reach. Particularly in the early decades, foresters were far from the agriculture-school technicians they often became after World War II. They constituted—and saw themselves as—a transnational cadre of engineers. They were biological cognates to the civil engineers who laid down railroads, erected bridges, raised dams, and dug canals. Like mining engineers, they leaped from site to site, spanning continents. They circulated through European empires, observing fire with a comparative parallax. But like good proconsuls, they saw flame ultimately through an administrative prism. They studied fire because they had to administer lands on which it was abundant. The purpose was to lessen fire's predations and better control the indigenes who relied upon it. They saw fire as their technocratic colleagues did brigandage, outbreaks of cholera, and marauding packs of wild dogs.

Still, imperial foresters were the first to examine fire in a global setting. William Schlich summarized the experience of fire in the British Empire within his five-volume *Manual of Forestry*, first published in 1889. In 1938, Georges Kunholtz-Lordat distilled the knowledge sieved from the French *ecumene* in *La Terre incendiée*. Until World War II, forestry provided the general (and often the only) medium for exchange, a common currency for fire science.

This regime could not persist. Forestry's imperial reach receded as Europe shed its colonies. The price of political power was to be politicized. The life cycle of state-sponsored forestry lasted, globally, less than a century. Nations discovered other purposes for their reserved forest lands, and these interests clamored

for a say about how fire would be managed and hence about the themes
of fire research. Wilderness, nature reserves and parks, wildlife preservation,
biosphere reserves, simple recreation—all compelled forestry to share jurisdiction over those lands and to explain fire as something other than a danger to
be fought. Equally, various sciences found fire on their margins and sought to
incorporate it within their field of vision. New perspectives in ecology, concerns
over atmospheric contaminants and greenhouse gases, the discovery of historical fires dating back to the shattering extinction event dividing the Cretaceous
from the Tertiary, alarm over slashing and burning in tropical forests, such
inquiries allowed fire to burst out of forestry's smothering embrace.

Everything expanded—the scale of analysis, the size of field research, the
institutional matrix, the reach of disciplines. The old imperial network had
decayed, but it kept reappearing in new avatars such as the International Union
for the Conservation of Nature (IUCN), the Scientific Committee on Problems
of the Environment (SCOPE), the International Geosphere-Biosphere Program
(IGBP), the Man and Biosphere Program (MAB), the International Decade for
Natural Disaster Reduction, and other agencies of the United Nations. For
instance, SCOPE inspired a series of symposia and volumes that synthesized fire
ecology for several of the major fire biotas of the Earth. And IGBP hosted the
International Global Atmospheric Chemistry Program (IGAC), which sponsored
massive field campaigns that linked biomass burning and emissions in Brazil,
southern Africa, and Russia, bolstered by smaller national efforts in the United
States, eastern and western Africa, and Canada, the latter most spectacularly
through the International Crown Fire Modeling Experiment. An old economy
of fire research centered in forestry met a new economy of remote sensing,
global inventories, and high-tech computer simulations, against which it looked
dumpy and shopworn.

Yet forestry remained the overseer of most of the developed nations'
wildlands. By the 1990s, it found itself paradoxically arguing in favor of fire.
Most public-land foresters had converted to an "ecological" perception of fire,
recognized fire's legitimacy, and sought to reinstate it. They stood as practitioners, pleading for the value of fire, while the *arrivistes*, claiming the high ground
of science and environmentalism, urged the abolition of fire. Even in the
United States, foresters found themselves stubbornly defending the use of fire

against those who no longer had personal contact with it, who viewed fire only through virtual imagery, and who saw no distinction between burning forests and smoldering tobacco, for whom open fires were just so much secondhand smoke.

For an interesting emblem of this extraordinary reversal, consider the case of Sir Dietrich Brandis and Johann Goldammer. For nineteenth-century forest-ers, Brandis was a patriarchal figure, the architect of British forestry in India and hence throughout the empire. Interestingly, he had begun his academic studies as a botanist, fascinated especially with the tropics, then converted to forestry as a means of reconciling knowledge with practice. By contrast, the twentieth-century Goldammer trained as a forester, also fascinated with the tropics, before moving into ecology. In the 1980s, he became editor of the United Nations Food and Agriculture Organization's primary fire publication, *International Forest Fire News;* an active organizer and participant in the swath of fire research campaigns that sought to link burning with atmospheric changes; and the host of the Global Fire Monitoring Centre, the guiding genius behind the admittance of fire to the working groups of the Interagency Task Force of the United Nations Strategy for Disaster Reduction—all in all as important a figure in contemporary fire as was Brandis a century earlier. Their two careers neatly track the changing patterns of fire research.

Granted such tectonic shifts, now would seem an appropriate time to reassess the overall status of fire scholarship. Assume, as a thought experiment, that such a program exists. What questions might it pursue? How might the central fire shape research? The query is, at present, as hypothetical as Heracleitus's. But if such an entity did emerge, these are the themes I would select to inform its inquiry. Consider them as something between founding axioms and working agendas.

One: The Biology of Fire

The bulk of fire science—and virtually all that might be considered controlled experimentation—has focused on combustion and fire behavior. This reflects both scientific possibility (what can be brought into a lab or subjected to semicontrolled field plots) and sponsors' ambitions (to control fire). There is no reason to abandon this tradition; on the contrary, there are profound incentives to bolster it because the practical need remains, the science is possible and is exciting as science, and whatever one wishes to study of fire ecology will require a sure description of the instigating fires. Effects will not be comparable if the causes are not comparable. We need a firmer account of the indices for and physics of free-burning fire.

Yet we also need to broaden fire phenomenologically beyond issues of fire behavior. I would argue that fire studies might be centered, ideally, within a biological context. Again, this is not an invitation to scrap the legacy of largely forestry-based investigations for the purposes of fire control. Instead, it invites a complementary vision cored in the fact that fire exists only because of life. Fire is, literally, a creation of the living world, and this fact offers perhaps the basis for a general theory of fire as a biological by-product. What would a general theory of fire biology look like?

No one, to my knowledge, has asked. There are formidable problems, not least that it does not seem possible, even in principle, to scale the biology of fire in the same way one can scale combustion. All fire behavior builds from a common reaction, the oxidation of hydrocarbons. A fuel particle joins the combustion of many molecules. A landscape fire amasses the combustion of many particles. Some scaling laws come into effect as one begins to factor in such environmental variables as slope and wind and as the fuel array organizes itself in complex ways that behave differently than as autonomous elements. One can imagine à la Laplace that if one knew the precise arrangement of fuel particles and the value of each one's combustion parameters (e.g., caloric content, fuel moisture, surface-to-volume ratio, and so on), one might forecast the outcome of any fire. The burning of forested mountains becomes a vaster version of the burning of a pine needle. Of course, this is hopeless. Apart from sheer numbers, the system contains massive indeterminancy.

Yet it is possible for fire physics to project a single process over a wide range of dimensions in ways that are not true for fire biology. Respiration does not affect biology over a range of scales, from cells to organisms to landscapes, as combustion does fuel particles. The biology of combustion at a molecular level is not, in minute form, the same as the biology of burning tallgrass prairie or a million hectares of closed-canopy forest. Scaling involves qualitative changes, a hierarchy of complex systems; its history likewise requires a braid of nested narratives. Are there, however, some common processes or principles, as there are for fire behavior, that would allow one to create a (loosely) coherent biology of fire? Is there a fire equivalent of the rock cycle or hydrologic cycle? Is a grand unified theory of fire biology possible?

Probably not, but it might be a useful exercise to try, for the endeavor would help to clarify what aspects of fire we can understand on physical principles and what aspects require us to make appeals to other modes of inquiry. The fact remains, fire exists because of the biosphere. That reality ought to frame fire scholarship more than it does.

Two: Fire Ecology as a Historical Science

Ecology has more in common methodologically with geology than with physics. For both, there are processes that can be analyzed in labs—the effect of heat and pressure on shale; the shearing on sedimentary strata; the impact of water on metamorphosis; the effect of scorch on trees, of heat on seed germination, of smoke on pathogens; and so on. The reason a rock exists as it does, however, reflects its history, and so it is with a biota, its trees, forbs, microorganisms. For too long, ecology accepted mechanistic—deterministic—models based on disturbance and equilibrium, an imitation of physical systems in which a "disturbance" represented a "force" against which an ecosystem "reacts." The same force would inspire the same reactions. In this way, one could erect a causal science based on (putatively) repeatable processes.

The flaw is that the initial conditions are never identical. Almost never are all the parts known, and rarely do they remain constant. But the critical issue is their organization. The ecological relationships are more important than the effect of fire on any single piece or suite of pieces. With few exceptions, these relationships are so complex that a model of disturbance and "succession," however elaborated, fails because an event is rarely repeatable. Rather, the actual structure, hence the dynamics, of an ecosystem is fundamentally historical and idiographic. It reflects context and contingency. Ecological models have tended to be insufficiently sensitive to such nuances. Landscapes do not behave with laboratory rigor. It is likely that fire ecology will never be predictive beyond vague generalizations. Like economics and geology, it can explain the past, not forecast the future. Ecology is a historical science.

That may be as good as it gets. But such a reconfiguration brings with it a tremendous asset. It allows ecology to weave people into its explanatory tapestry, which leads to my third proposition.

Three: Humanity as a Fire Agent

However much researchers (Americans, especially) might long for a pristine fire ecology uncontaminated by humanity, it does not exist in any meaningful form. The geography of fire on Earth looks as it does because modern humans have held fire as a species monopoly since our origins. One reaction is to lament this fact and to seek to strip away those blemishing behaviors, creating landscapes like Ansel Adams photographs void of the human presence. Often scientists have attempted intellectually to achieve, in effect, the same goals the public-land agencies have pursued: to exclude anthropogenic fire from the field.

Increasingly, this strategy has become meaningless as well as counterproductive. As long as people have lived in a place, they have shaped its fire regimes. It makes more sense to explore in a systematic way just how they do so. Rather than base fire scholarship on natural fire, which exists on any scale largely because people have chosen to create "natural" habitats for it, like public wildlands, a more practical and conceptually tractable strategy is to found the study of fire on its primary agency, people. It makes more sense to consider wilderness fire as a special case of anthropogenic fire than to analyze anthropogenic fire as a muddled variant of natural fire. So the question arises, Is there a conceptual anthropology of fire?

Remarkably, the subject has been studied little. Omer Stewart announced a bold beginning in the early 1950s, which Henry Lewis later elaborated and a handful of Australian anthropologists have explored further; more recently Richard Guyette and Daniel Dey have sought to model the process in select American landscapes. A literature has bubbled up mostly concerning whether, at the time of European contact, indigenous people burned and what changes

European settlement induced. Some of this is substantial and thorough—
Australia is particularly well endowed. But most studies are absorbed within
the general discourse over contact, not fire studies. They are not plumbed for
evidence that can lead to an understanding of humanity as a fire agent. There is
no *fire* bond between aboriginal foraging, for example, and swidden or between
big-game hunting and industrial forestry. How do people use or exclude fire?
What is humanity's utilitarian relationship to fire? Are there general processes
and patterns that span environments and cultures?

Fire is an amorphous, plural technology. A systematic investigation into
its biotechnological character and varied uses has hardly begun. This confusion
over fire technology often muddies arguments over how anthropogenic fire
behaves ecologically or screens off an appreciation for fire's role if it appears as
an all-purpose catalyst for human fiddling with the environment rather than in
the simple guise of an implement. But a perhaps deeper muddle results from
the inability to integrate ideas and institutions into the dynamics of fire ecology.
Because humanity is such a fire presence on Earth—directly fidgeting with ignition, indirectly tweaking fuels—an ecological explanation should incorporate
the means by which we apply our firepower. Why do we act toward fire in
certain ways? How do we impress those understandings and visions onto
landscapes?

The brief answer is, we conceive of fire on the basis of our information
about it, and we act through social institutions. What we know (or don't know
or think we know, wrongly) matters as much to fire's ecology as the moisture
content of fuels. How we move knowledge through institutions affects fire's
ecology as fully as the turning of the seasons. The flow of knowledge is as vital
as the flow of nitrogen or sulfur; the structure of institutions has molded biotas
as surely as mountains and rivers and the rhythms of the seasons. Scientific
periodicals, professional journals, books, popular magazines, television—all
have packaged and shunted the information upon which society decides how it
proposes to manage fire. Thus, what happens at a fire in California can affect
fire practices in Georgia. A fire in Australia can influence fire programs in
South Africa. The Yellowstone National Park fires of 1988 shut down prescribed
natural-fire programs throughout the United States and gave pause to fire

strategists around the world. Fire ecology, in brief, cannot be understood without reference to fire culture.

Four: Industrial Combustion as a New Species of Fire, One That Complements and Competes with Other Forms of Earthly Combustion

Perhaps because we traditionally have chosen to consider fire ecology within natural landscapes, we have not explored the ecology of what has become the dominant form of combustion on the planet (bar only cellular respiration): the burning of fossil biomass, which is, for fire history, the meaning of industrialization. An estimated 60 percent of earthly fire, as measured by effluents, now results from industrial burning. The subject typically resides with mechanical engineering in its quest for better combustion chambers and fuel-oxygen mixtures or with certain efforts to model the mobilization of carbon or the climatic consequences of massively pumping greenhouse gases into the atmosphere. Yet almost no serious inquiry has explored how industrial combustion might connect with or compete with traditional fire practices and hence how it reshapes fire regimes. What is industrial fire's ecology?

This is where robust fire studies might show their conceptual power. The burning of fossil biomass seems simple enough: it seems, indeed, a problem of techniques. In fact, a case can be made that the act marks as profound a shift in planetary fire history as the human ability to kindle sparks or to create fuel out of living biomass. Industrial fire is reshaping every fire regime on the planet, and not through atmospheric loading alone, but by substituting for tool flame and tame flame, by redefining human land use, by rearranging the geography of living fuels, by forcing conceptions of ourselves and our relationship to the world around us.

Little remains untouched. Industrial fire has remade the ecology of urban fire by virtually banning open flame altogether; the ecology of agriculture by substituting fossil fallow for living fallow and thus again erasing open flame; the ecology of public wildlands by leaving them for recreational or preservationist purposes and by shunning anthropogenic fire in favor of natural fire. Initially, fire suppression became more effective because the force of internal combustion could go head-to-head against the force of free-burning fire. Even backfires have receded as a firefighting technique in favor of aircraft, mechanical pumps, chain saws, crews whisked about by vehicles, and flame-retarding chemicals.

Today the Earth is dividing into two great combustion realms, one devoted to the burning of fossil biomass, one to living biomass. But how, exactly, has this immense fissioning occurred? What are its ecological consequences? How might it be possible to manage this pyric transition? And who should study it?

Five: The Study of Fire's Frontiers

Two realms of combustion, yes, but three variants of fire, all of which continue. The Earth's long burning has two critical frontiers. At one border stands natural fire; at the other, industrial fire. Between them ranges anthropogenic fire or that species of combustion that relies on humans to create spark or fashion fuel and that burns living biomass. Those fire frontiers matter conceptually because they set boundaries between and hence help define the different varieties of fire. They matter practically because the two most lively areas of fire research track their contours. Neither frontier, however, has undergone systematic study.

The enthusiasm for nature reserves has created a need to understand how "natural"-fire regimes differ from anthropogenic. Over and over again, the desire to preserve natural landscapes as parks, wildlife refuges, or wilderness clashes with the fact that, at least with regard to fire, many of these areas were cultural landscapes. The firestick, not the lightning bolt, defined the fundamental matrix of fire. Left to itself, a natural-fire regime might well create a landscape unlike that which was set aside for protection, if it is even possible to reinstate such a regime without constant finagling by humans. Regardless, the query has prompted a considerable outwash of studies. What they lack is a coherent context in which they might be compared as fire scholarship and some principles condensed from the brew.

The frontier between industrial fire and anthropogenic fire is even more vexing because so much of the influence is indirect—as a result of perceptions held by urban populations and applied through governmental agencies, for example—and because all three fires join the swirl. Natural fire, anthropogenic fire, and industrial fire would seem to form a pyric three-body problem without an exact solution. Yet the "crisis" that often follows the removal of anthropogenic fire is, in truth, generally an outcome of industrial fire; and it is the propagation of industrial burning that is remaking the planet. How this "frontier" actually works is an issue of immense significance to the world.

Other frontiers exist, of course. Within the realm of anthropogenic fire, various peoples relate to fire variously and arrange it on the land according to their talents and desires. Farming exploits fire differently than hunting, pastoralism differently than farming, tourism than foraging, and so on. At present, there is too much lumping and splitting to sort out the various contributing causes to a shift in fire regimes—how much weight to give natural causes, how much to humanity variously equipped with technologies and economic institutions, how much to industrialization. What, for example, passes as a "European" influence may represent a shift in fire practices from hunting to farming or the impact of imperialism in which practices cultivated at one site are thrust into another or the ponderous and diffuse tentacles of industrialism rearranging fuels by the way it encourages fire to be applied, withheld, or reorganized. Here it might be wiser to split what has been lumped under the "European" label. Elsewhere one might better lump under "anthropogenic" fire all those practices that must stand, awkwardly perhaps, against natural or industrial fire.

Six: The Political Ecology of Fire

Granted that so much fire research—nearly all, in fact—has had government sponsorship, it is remarkable that so little is known about the politics of fire. Examples of fire politics exist from history, and they especially abound today because of the rash of new nations created from the breakup of the Soviet Union and its satellites, from newly democratizing countries, and from the still unsettled decolonization of Africa. The opportunity exists to study in some systematic way how fire practices are shaped at a national level, how fire policies are formed, and how fire research is sustained.

The pragmatic consequences are real. What fire policies and programs should Kazakhstan adopt? What, realistically, might Indonesia do about the smoke that burning Borneo casts about the region? How might Russia negotiate a soft landing for its once-mighty aerial firefighting service? How should Ghana respond to rural fire menaces? (There is, to my knowledge, no evidence of a successful fire campaign against rural fire. The "solution" comes from removing the rural population, either through eviction, collectivization, or urbanization. But no one has studied systematically the political context of adapting fire practices to industrialization or decolonization, much less to both together.) Not since the high tide of European imperialism has there been a better opportunity to shape how humanity should apply and withhold fire. What policies are suitable? When do policies really matter? When are research programs simply exercises in political misdirection or public-works programs for scientists? How do we create (or fundamentally realign) political institutions for fire management and fire research?

Seven: Humanity as Keeper of the Planetary Flame

Fire may be the most distinctive index of ourselves as ecological agents, the one thing we do that no other creature does. It measures—testifies to—our unique presence. The burden of determining how to use fire, however, falls to culture; fire practices are, ultimately, a moral matter, relating to who we are and how we should behave, and for this reason a robust agenda of fire studies must involve the humanities. Any honest rendering of fire ecology should examine, at some point, the culture of fire.

The prevailing perception of fire is that it is, obviously, an object of nature and hence a subject for natural science. Nothing could be less simple. Fire is a hybrid, as we are, of nature and culture. Almost nowhere does fire appear without the human hand and mind somewhere in evidence. So fire practices are not the outcome of some autonomic response, the reflex of instinct, but they reflect choices people make about what fires to introduce, where to withhold flame, when to apply the torch, and why one might reshape landscapes. Those choices differ by cultures. We can analyze a society's fire practices as we can its literature, architecture, or legal institutions. They reflect each peoples' view of themselves and how, in reality, those views appear on the ground. Our fire practices are not only a means of ecological agency but of cultural analysis.

an exchange for all things?

Actually, one of the founders of folklore, Sir James Frazer, had written incisively about European fire ceremonies and global fire myths in *Balder the Beautiful, The Golden Bough,* and *Myths of the Origin of Fire.* Almost nothing has followed, save Herbert Freudenthal's *Das Feuer im deutschen Glauben und Brauch,* Johan Goudsblom's *Fire and Civilization,* and, in somewhat different ways, Claude Levi Strauss's *The Raw and the Cooked* and Cristina Garcia Rodero's *Festivals and Rituals of Spain.* The candle of fire-cultural studies has burned very low, indeed.

Even to friendly critics, my proposed agenda for fire studies will look suspiciously like my own scholarship—and may be dismissed as a disguised apologia. (Usually I try to stay one continent ahead of my critics.) But Heracleitus's challenge is not going away.

Creating a formal discipline of fire studies is probably a bad idea. Probably we are better served by errant scholars who stray across broken disciplinary fences or the occasional eccentric who extracts fire from a host of fields, like Vladimir Nabokov netting butterflies, and then mounts these representations into a common display. (If a formal organ emerged to study fire, academics, in particular, likely would waste energy and goodwill trying to define and defend the field rather than using fire to explain other phenomena.) Perhaps fire's nature is to thrive as a catalyst for other scholarship. Probably we ought to approach the big questions of fire scholarship as we do big fires, through an indirect attack. But maybe fire has something to offer in and of itself and like a campfire in a dark night can serve to muster round it a gathering of far-searching scholars.

proscribed
burning
2

Biosphere 2 has a rainforest, a small sea and coral reef, a savanna, two man-
grove marshes (one fresh, one salt), microbial producers and decomposers,
soils, a self-contained atmosphere, rivers and rain—it holds a biosphere under
glass. What Kew Gardens was to imperial Britain, Biosphere 2 is to global
Greenery. It seemingly has everything a miniature Earth should have. But it
does not have fire. Why?

The lapse should seem odd. It would appear an omission so fundamental
it saps the very premises underlying what is a kind of biotic Mall of America,
Big Ecology's theme park. After all, fire takes apart what photosynthesis puts
together. It is respiration of another sort—a fast combustion outside the cell to
match the slow combustion within. Fire has been on the planet since terrestrial
plants took root; fossil charcoal dates from the Devonian. Adaptations to fire
are rife and robust. It is considered basic to the ecology of many biotas. Either
fire is fundamental and Biosphere 2 is deeply flawed, or it is optional and
our contemporary understanding of fire ecology, at least in America, is an
intellectual construct as contrived as this exotic greenhouse. How can it be
absent?

For lots of reasons, but two stand out. One, there is no source of ignition—no volcanoes, no cherty stones toppling down slopes, no lightning blast-
ing snags out of a glassy sky. (Lightning outside is feared; the Bubble bristles
with lightning rods.) Above all, the place lacks hominids avidly kindling fire
to all sides. The people who inhabit this mockup of Spaceship Earth are few
and by far the largest creatures in its realm. This elicits the second, more basic
question, the matter of scale.

On a molecular level, combustion is immutable. Cellular respiration is
identical, cell by cell, oxidized molecule by oxidized molecule. The combustion
process is already miniaturized: the Krebs cycle fits into the mitochondria like
microchips in a computer. Bigger organisms add up combustion increment by

increment. There is no real scaling such that combustion itself undergoes a change in kind. A giraffe does not respire differently than a bacterium. But free-burning fire is different. It is relatively unconstrained, bounded only by terrain, airshed, and biomass. It may be small or huge, a single pine needle or a crown fire of fifty thousand acres. True, the latter is a massing of the former: the crown fire burns particle by particle. Yet that massing of flaming needles has a character greater than the sum of particles consumed because they burn together, not sequentially, and thus merge their heat. In a cell, the environment for combustion is invariant. It comes and goes, and the mitochondria remain unchanged. In an ecosystem, the environment shapes combustion in profound ways, and combustion returns that compliment by restructuring those environs. Free-burning fire can range in size from a match head to a mountainside. Its potential scale argues against its biospheric presence: a solitary wildfire could swallow all of Biosphere 2 in a single surge.

One doesn't need apocalyptic flame, however. The chemistry of free-burning combustion is enough of a threat. Even a small fire can gulp vast amounts of oxygen. Every photosynthesized sugar molecule demands, as its tithe of combustion, two atoms of oxygen for every molecule of carbon it contains. Burning a modest patch of Biosphere 2 woodland would suck the dome dry of oxygen like a candle under a bell jar. Rough calculations reckon that the oxygen would vanish in eight minutes. Emergency evacuations plan for four. (Axes line corridors with instructions to break glass in case of fire—the glass of the Bubble itself.) Thus, Biosphere 2 doesn't have flame for the same reason it doesn't have elephants and whales: it isn't big enough. Granted the proportion of Biosphere 2 to Earth, the dome might accommodate perhaps a birthday candle or two every year. And that is one reason why people don't burn there—why a species that has set fires for all of its existence in every imaginable environment declines to do so in an edifice intended to emulate Earth's environment.

What does this say about fire on Earth? It says that cellular respiration far transcends free-burning fire in the calculus of planetary combustion. However spectacular, fire flares only sporadically, at most by season and in select landscapes. A number of biotas thrive without it, and others tolerate it, accepting it but not demanding its regular return. In the grand cycling of earthly chemicals, its contribution is minor compared to the exchanges wrought by cellular combustion and the burial and exhumation of carbon.

It says that size matters, that free-burning fire is, in fact, a phenomenon of scale. It exists today in a great geographic warp between two worlds, between a cellular-scale combustion based on the individual oxidation of molecules and an immense industrial-scale combustion based on special chambers to burn fossil biomass. The latter is an uncanny echo of the former. As much as molecular biology, it parses chunky hydrocarbons into their combustion-active parts and refines the air into its chemically vital ingredient, oxygen. There is no more opportunity for fire to escape from the one than from the other. They are, in truth, species of combustion that are hardly recognizable as fire. Between them, however, they frame the realm of flame.

Fire in its vernacular sense exists in a vast, largely unconfined cauldron in which the active agents are diffused among earth, air, and life. Fire relies on that immense sloppiness. If its context were confined biologically more precisely, if the tongues and grooves of its ecological carpentry fit with micrometer exactness, fire could not exist. There would be no space for the eruptive reaction to occur, no residual fuel unconsumed by biological agents on which the flames could feed. Instead, fire depends on slopes and gullies for combustion chambers. It responds to wind and layered air masses and must abstract that fraction of the airshed, oxygen, that it requires. It searches out combustibles from among the biomass stored in soils and stacked on the surface, only a sliver of which are normally available to burn. For these reasons, fire is spotty, its occurrence lumpy, its behavior full of spunk and whim. It knows limits through a biota that, by means of evolution and ecology, can contain its potential explosiveness by not allowing flame to gorge uncontrollably on the stored biomass around it. This is a system of generous checks and balances, with lots of flex, patchiness, and wandering.

Biosphere 2 has no space to so manifest surprise and slack, no ecological wriggle room. Its scale is too small, its dimensions too tiny, its history too short-lived. The more interesting question is whether, were it possible to expand the Bubble to the scale of the Sierra Nevada, Biospherians would wish for fire or whether they would hew to its Gaia-like premise that the living world buffers against just such biotic eruptions, that a congested Spaceship Earth no longer can tolerate violent and wasteful outbursts, that fire is either a parasitical phenomenon or an epiphenomenon that, like lichens on larch, can be scraped away from the thick bark of the biosphere without harm. It no more belongs on Biosphere 2 than a fire on space station Freedom.

What the specter of a flame-free Biosphere 2 tells us is that free-burning fire exists not primarily as an engine of planetary biogeochemical exchanges, but as a catalyst. It exists precisely because the real world is not a machine, engineered to exact specifications. It is not a clock, not a greenhouse, but a fermenting, crawling concoction that allows fire—and not infrequently demands it—to unclog and spark its peculiar and often unpredictable biochemistry. It often needs an occasional flame to weld its loose parts together. If all the pieces worked with unflagging completeness, there would be no need for new species or new ecological arrangements. The Great Chain of Being would unfold full and immutable. It does not. Fire thus exists for the same reason that evolution by natural selection does. The untidiness that allows for change also allows for (and sometimes demands) fire.

The place is not merely an experiment: it is a model of nature made real. In a world eager for a biocentric philosophy, it stands as a Crystal Palace, and, like its Victorian antecedent, it proposes a kind of technological utopia—this through a biological medium rather than a mechanical technology of iron and steam. The gleaming glass dome is an industrial garden: a city made of biospheric components, constructed of cellulose and lignin rather than concrete and steel. It substitutes greenhouse glass for greenhouse gases. It exists only because a culture could imagine it and wish for it.

That such a culture could conjure up a model world that gleefully banishes fire should give pause to fire fundamentalists, both those who believe that fire ought to be abolished because it is bad or useless and those who insist that fire can never be erased because it is good or essential. It is not enough to chant that fire is natural. Biosphere 2 suggests the debate is not about biotheological axioms and immutable principles but about scale and values. A habitat informed by genetic engineering and Gaia gardens does not need fire.

A habitat of mountains, Big Tree sequoias, and tallgrass prairies bending in the wind does. Clearly, it is possible to make landscapes from which fire can be eliminated, at least in principle. That, in fact, is the goal of the industrial city, and across environments ranging from temperate Europe to boreal Canada to mediterranean and subtropical America such urban enclaves have become a reality.

Biosphere 2 shows that this vision can be extended to other environs as well, and it implies that as modern biotechnology becomes as robust as modern physical technology, whole landscapes can be remade and no doubt will be. The old domestication of the Earth through farms and pastures will evolve into a high-tech reconstruction as little resembling swidden plots and kitchen gardens as the ancient Roman agora does modern Manhattan. Enjoying a quantum leap will be Cicero's observation that people, through their arts, have made a second nature out of original nature. Will the inhabitants of such a world want fire or not?

Biosphere 2 suggests they will not or suggests powerfully that humanity may prefer to have its combustion channeled through power lines or housed in genetically supercharged decomposers without the bother of smoke, soot, and ash or the general untidiness of flame, to say nothing of fire's casual orneriness and indifference to human ambition. Yet it would be an odd vision: that the one species to enjoy a monopoly over fire should strive for a world in which that firepower would become unnecessary and undesirable.

It is not so easy to segregate humanity from flame. A world without fire is a world without need of a fire broker. A world without the slack to accommodate fire is a world with no use for a creature uniquely equipped to handle fire and to track down or fashion niches in which it can flourish. There is no place for a clumsy, opportunistic, impulsive, clever, teachable, willfully disruptive creature like ourselves. In fashioning a world without fire, the Biospherians perhaps unwittingly have fashioned a utopian world without environmental ills, and that means they have abolished humanity, save as a general machinist, scholastic commentator, and tourist of its miniature marvels.

What Biosphere 2 proposes for humanity is a role rather like the Deist notion of God: an Author, a Creator, a grand Clockmaker, who wrote the script, erected the scaffolding, and left the machinery to run by itself. An ideal world—and Biosphere 2 is nothing if not an ideal—has no role for humanity as an ecological agent, it would seem. The abolition of fire almost demands the abolition of Earth's unique fire creature. For a species committed for all its evolutionary existence to active intervention through controlled combustion, it is a decidedly puzzling notion, like a self-deconstructing text beloved of postmodernists. But the future of fire belongs with those competing visions, whether they be a model or a mirage.

The aptly named nearby town of Oracle shows the two extremes for fire. There is the gleaming glass *domus* of Biosphere 2, the very image of an urban peoples' vision of a benevolent nature and sustainable habitat. It is a site that so controls ecological processes—that devises biotechnological surrogates for them—that it can banish fire and, in fact, must banish it. There is no more need for open burning than in downtown Tucson. Beyond the dome lies an alternative universe. Rippling through the crumpled terrain of the Sonoran Desert and up through the sky-island mountains and to the Mogollon Rim, a riot of monsoon thunderstorms explode, kindling more fires with greater abandon than anywhere else in North America. Neither world will drive out the other. How they interact, how each will shape humanity's understanding, will determine what kinds of fires will thrive in what proportions.

Where fire finally fits will depend on scale. Even the built environment can be too big to obliterate all the possible niches for fire, and no doubt glitches in the social software will encourage some outbreaks. But such fires as range over the general landscape may resemble the eccentric fires of modern cities, restricted to unused warehouses, run-down tenements, abandoned lots—precisely those portions of the cityscape not yet brought under full control. In principle, if at some cost, these fires can be wrestled into extinction.

The more interesting question is whether, based on values, the people who inhabit such a world will want some fire, much as industrial peoples want wilderness. Leaving wildfires with grizzlies and pileated woodpeckers, however, is a very different order of fire management than that which has shaped humanity since the Pleistocene. It reduces the basic discussion to one of nature protection, a splattering of insulated places, not to one of fire practices as a defining act of human character that has reached every place people have. It leaves wildlands as a kind of inverted Biosphere 2, an undomed patch in relation to which humans remain, again, as spectators. Free-burning fire would flourish only in reserves as contrived as the fire-free Biosphere 2.

There is more at stake than fire as a symbol, although it is certainly that—synecdoche for a more expansive vision of our relationship to Earth. Under the searing Sonoran Desert sun, two visions of a greenhouse Earth flourish—one under glass, the other charged with light-absorbing gases. For both, combustion is fundamental. For the first, though, there is zero tolerance for flame, and the prospect of a Biospherian wielding a firestick is as improbable as one wielding a harpoon. For the second, flame is inevitable, and the proper response is to accept it in certain places and to control it in others, largely by deliberate burning. The visions are competitive but not exclusive: they can coexist except at specific sites. But they are likely to become more exclusive with each passing decade.

We cannot separate our visions of the world from how we act in it. What we do with fire is, by and large, what we will do with the Earth overall. Our choice regarding fire inevitably will involve a choice about what kind of world we wish to inhabit, which is to say, what kind of creature we wish to be.

firebug

The general inability of modern humans to get the right kind of fire in the right place at the right time is so universal that a cynic might be tempted to observe that nature would do well to give the task to some other creature. This is unlikely to happen. Evolution is too slow, and *Homo sapiens,* however slack its oversight, would stubbornly resist the surrender of its unique power. But today we don't have to wait for evolution: in principle, we can fashion such a beast through genetic engineering. Moreover, by inventing a servant species to do a job that we apparently regard as the ecological equivalent of garbage collection and ditch digging, we can outsource the task yet retain nominal control over our fire monopoly. We can program the organism to do what we should do but are too sloven to achieve. What sort of designer species would do? What would make an ideal fire creature?

The basic problem is to match flame with fuel. This task requires the ability to produce ignition and combustibles on demand, the capacity to recognize when and where to kindle a fire, and the will to set it when the occasion merits. The near-universal failure of people to recognize what needs to be done and then do it recommends that these instructions be coded by instinct, and ease of biotic maintenance as well as simplicity recommend for an insect. The ideal creature would be a firebug.

How can this creature accomplish the first task, to set fires? The most direct way is to carry a brand (or ember) from an existing fire, rather as early hominids did. This is simpler than it sounds. The Philippine tarsier will pick up coals, and wedge-tailed Australian eagles can grasp burning debris in their talons and deposit it elsewhere. The mechanics of grabbing and carrying firebrands are possible: insects already pick up things. They would need either insulated hands to hold hot debris or be coded to wrap up the burning material in a leaf or nutshell in order to carry it. None of this is impossible.

The difficulty is that a source fire is not always around when you need it. The firebug would have to maintain one, a kind of insectivorous vestal hearth, to ensure that a brand would be available when required. This argues for a social insect, with some caste feeding the fire much as ants and bees divide up the task of nurturing their young. Even so, there are limits to how far one can carry a brand. If small, it will expire rather quickly, and if large, it will prove difficult and perhaps dangerous to haul. The whole apparatus to support an intricate society of firebugs suggests that the genetics will be complicated. The easier solution is to equip each individual—or significant fractions of the population—with the ability to move through the terrain and to start fires at will. The firebug should be able to wander widely, ideally by flying, and it should have the capacity to kindle fires on the spot.

Is this possible? Of course. Lots of insects fly, including big ones. None start fires, but various crickets and grasshoppers rub their legs rapidly to produce sounds and these movements could be tweaked into a biotic fire drill. The heat only needs something to act upon. So the firebug must collect kindling, which is no different in principle from gathering select substances to make nests. It would also help to have an accelerant to lower the kindling temperature, which the firebug could produce, much as insects do digestive juices. By inserting the proper bacteria into its gut, the firebug probably could concoct an alcoholic brew, which it then could excrete over the assembled tinder, or a cocktail of volatiles that might ignite more or less spontaneously. On a hot day, the mix easily could boil over into flame.

Whether the firebug itself could escape the flare-up is doubtful, but from the perspective of their human masters, this is hardly a limiting consideration. In nature, ritual self-destruction could prove fatal for the species; any fire creature would have to ensure either that it escaped or that the pyre contributed to its reproductive advantage. Thus, a natural firebug might kindle fires only after birthing, or it conceivably might find an advantage in the flames, which could, for example, assist the survival of its very young by driving off competitors and predators. But such considerations will only encumber the design of a servant species—really, a flying match—whose self-immolation may be the purpose of its existence and whose propagation would depend on biotech labs.

The ability to kindle *fires*, as distinct from *ignition*, requires that the flame strike the proper fuel. The flame has to spread, which means it must connect with combustibles—that is to say, biomass in amounts and arrangements that make them available for burning. Most biomass is not so available, or if it does burn regularly, it does so in spasms as cycles of wetting and drying make portions of it available. The firebug will have to recognize fuel when it sees it. Again, this is not insurmountable: creatures do this already. They adjust their behavior by season and know what of the world around them is food and what isn't. A happy side-benefit is that a properly timed fire will defang lightning as a competitive source of ignition. The firebug simply will burn when suitable fuel is on hand, well before dry lightning might appear on the horizon.

Still, fuel foraging presents a problem. There are many kinds of fuel, an infinite variety of settings, an almost unbounded set of possible combinations to sort through. No insect could hope to contain the programming to recognize them all. Instead of one firebug, there would have to be a multitude. Firebug would be a genus (at least), not a species. Thus, the difficulties begin piling up. The escalating requirements dampen the attraction of an instinct-driven creature, unless one is prepared to create hard-wired firebugs for all the fire regimes of Earth. There would be as many firebugs as ants or locusts. The alternative would be to endow the genus firebug with the capacity not only to kindle spark but to create fuel. Again, plenty of insects do this already—cutting or desiccating and thus inadvertently readying plants for burning. It should be possible to genetically train them to do it in order to burn deliberately.

From the perspective of fire, this combination holds magnificent attractions. The opportunities to get flame and fuel together multiply like weeds. Not only could fire routinely pass through the biotas that have adapted to it, but it would be possible to expand fire into nearly every imaginable landscape. Only the most hostile places could evade fire. The firebug could truly become a pyric clone of humanity. All the places that humanity has touched with fire it could as well. But the escalation of possible acts comes with an escalation in complexity.

It is difficult to imagine an insect that can both gather fuel and set it afire. Even a social insect, a colony in which fire starters formed one caste and fuel gatherers another, would be tough to engineer. The very specificity of the programmed instincts would demand an equal specificity of creatures. One would have to rely on a plentitude of species, perhaps a plentitude of genera. This magnitude of eco-engineering is beginning to look like work—far too difficult for nature, which has had half a billion years to work out the genetic equations and has declined. The integration of all these firebugs would introduce still further complications. The more precisely a firebug would do its job, the more likely it is that breakdowns would occur. Moreover, all this assumes that, once released, the creature would do our bidding. We would need to build in mechanisms to contain it, to ensure it did not become a rival rather than a robot. Instead of controlling fire, we would need to control the firebug. The cost of contriving and then containing such a creature easily would exceed its value as a device to slough off our inherited duties as keepers of the planetary flame.

The only real way around this proliferating complexity is to craft a creature that can be taught. It could learn, and through education it could be capable of adjusting its fire starting and fuel stashing to the Earth's endlessly varied circumstances. Such a schema, however, moves us beyond insects. In the end, it only substitutes one set of instructions, learned, for another, instinctual. It is, in fact, what evolution decided to do when it conveyed to *Homo sapiens* responsibility for fire. The end result would be a designed species rather like ourselves.

So the good news is that *H. sapiens* may be the best that evolution by natural selection can do. Nature has the genes more or less right. Having eliminated any competitors, it is difficult to see how, other than by artificial means, any alternative might appear. We don't need to create a biotic surrogate because the outcome, in all likelihood, would look like us. What we need to mend is the culture that, in place of genes, guides our behavior.

And that is the bad news. The problem is us. With respect to fire, industrial societies have forgotten more than we've learned. We're dumb and getting dumber. We prefer the virtual fires of TV and the gluttonous combustion chambers of SUVs. We've turned our oldest ecological ally into an enemy. We've tried to give back to nature what it gave, unreservedly, to us. If we did try bioengineering, we probably would bungle the genetics and get an insect that would suppress good fires and start bad ones. We might end up inventing a worse example of ourselves.

There is no need, however, for heroic measures. Fire looks problematic only because we have forgotten what it means to us. So although it is clear that Earth will survive in the absence of widespread combustion, it is a more interesting world with flame and a world that burns with enough regularity that fire may be listed as among its defining properties. Ours is uniquely a fire planet that hosts a unique fire creature. We may have forgotten that fact, but nature has not, and some day we may startle awake to discover that, against all odds, another species has emerged that also can manipulate fire, that, although it cannot do everything we can do, can do some basics and through rapid selection or long apprenticeship can learn the rest. If that moment ever comes, our world will end in the flames we wrongly feared and foolishly neglected.

fire's

new

new

thing

Alpine Village East (AVE) is a recent forest subdivision south of the unincor-
porated village of Alpine, Arizona, whose flanks brush against the Apache
National Forest. It demonstrates both the reality of the intermix fire scene and
why it may be time to look ahead to the next problem fire. When AVE was platted
in the late 1980s, its covenants, codes, and regulations specified a "natural"
look—wooden (or slate) roofs, compact housing with lots of woodlands left, few
trappings of city life. The original scheme included a ski resort, which never
happened. When the speculative frenzy, financed by looted savings-and-loan
companies, folded, the project reverted to the Resolution Trust Corporation. In
1994, it came out of Resolution hands and joined the boom of the Nasdaq
Nineties. Within five years, it had a county road system; new housing sported
metal roofing; residents raked their "yards" of pine needles within an inch
of their wooded lives; Forest Service fire officers offered free inspections and
commenced a major cleanup along the boundary; the streets had speed limit
signs and yellow hydrants; and Alpine had created a fire district and staffed
a volunteer fire department. As lots filled, the landscaping began to look as
much like the White Mountain forest as the "desert landscaping" of Phoenix
and Tucson does the Sonoran Desert. The principal fire hazards in AVE lay with
the dappled texture of its absentee ownership. The threat remained real, but fixable. That, in fact, had always been the case with the intermix fire: it had been a
witless problem to have because it was amenable to technical solutions. Within
another decade, AVE likely will advertise a fire hazard not much worse than
Alpine itself, which is to say, well within the purview of small-town America.
Within a decade, the American fire community can anticipate that the intermix

fire problem will fade away from dominance. Within four or five years, it should be possible to sketch the likely contours of a successor. What might that be?

Alpine's setting offers a clue. The place has long held a fabled wildlife, which Aldo Leopold, then a young ranger, immortalized in two stories. He told how, on the slopes of Mount Escudilla, which looms over Alpine to the north, Arizona's last grizzly, Old Bigfoot, died from a shotgun blast from a trip wire set by a government trapper. Leopold described how, somewhere to the south—the exact location ambiguous and vaguely allegorical—he shot a wolf and then watched the "green fire dying in her eyes." It was a slow epiphany, an insight into what he termed "thinking like a mountain," and it became the basis for perhaps the most famous wilderness essay in twentieth-century America. An ecological ethos required one to think in a different calculus than that which proposed that fewer wolves meant more deer.

The two episodes were of a piece. Bigfoot was killed to make the world safe for cows. The wolf was killed to make the world safe for deer. So it was that fire control would make a world safe for trees. Early rangers and conservationists hunted down fires with the same hard ruthlessness that they pursued bear, wolves, coyotes, and other vermin. Fewer fires meant more forest. Fire was as intrinsically evil as the wolf, and in fact European folklore and early fire-prevention posters often linked them explicitly. Free-burning flame was likened to a pack of "fire wolves," both equally bloodthirsty and both justly subject to government bounty.

Times change. Not least thanks to Leopold's essay, the wolf has become a symbol of a wild America in need of preservation. So it has been with fire. If hunting wolves was wrong, so was suppressing natural fire. If the "mountain"—the wild—needed its wolves in order to work rightly, so it also needed its fires. Campaigns to reintroduce both wolves and fire cranked up and became unstoppable. The wolf returned to Yellowstone; so did big fires. The Mexican gray wolf was reinstated to the White Mountains, and so were fires. The nation's first primitive area—one set up at Leopold's instigation—the Gila, lies a bit to the southeast of the White Mountains, a lower, drier range, not capped with lava and volcanoes. Here began one of the first experiments to tolerate lightning-set fires. A lightly populated environs, a fire-prone landscape, a remoteness that would allow flame and smoke to roam and blow more or less where they would—with little fanfare, fire made a hesitant return. The domain of natural fires expanded into the White Mountains and their shaggy flanks. By 2000, the realm of reintroduced fire and wolf roughly coincided.

There were many more fires than wolves, which made wolf reintroduction simpler. Moreover, the potential for a fire to blow up was greater than the likelihood that the wolves suddenly might multiply and mass into packs and turn killer. Prescriptions existed to decide when one became a threat, when a fire might go feral, when a wolf became dangerous because it attacked cattle or threatened people, when, in brief, they had to be hunted down and shot. One of the reintroduced wolves, in fact, did wander into Alpine and past children waiting for a school bus. A large natural fire, still "under prescription," cast a smoke pall over the Alpine valley for nearly a week.

The larger lesson, however, was that success depended on the setting. It was not enough to loose bewildered bear, wolves, and flame back into the scene. The problem was not simply the wolf but its potential habitat and whether it was enough to sustain and contain a wide-ranging creature. The wolves would prove only as viable as the lands they roamed. So it was with fire. Free-burning fire would thrive only as its habitat allowed. The early reintroductions occurred in places so remote that they could absorb many miscalculations. But as their domain has moved closer to settlements, the chances for bad encounters have multiplied. That intermediate realm, however, is where the action lies. Wilderness has had its day as the poster child for American fire. More recently, the intermix has replaced it; that too will soon fade. What remains is the vast dominion between them. These are lands seemingly without special values, but they constitute a realm often badly mauled by a century of hard use, of logging, grazing, predator extinction, and fire exclusion, a place as harmed by ecological sins of omission as of commission. It is where the American fire community next needs to go, and AVE's experience shows perhaps how that might happen.

The working dogma of the intermix era has been "defensible space." Its definition is more or less self-evident. It means that, after ensuring that a house is not plastered with combustibles, one looks to its immediate surroundings. One culls the trees, cuts back the scrub, rakes away debris that brush against wooden siding and decks. The size of the space varies because a fire can propagate by leaping as well as sweeping. Flames can rush ahead of a burning source, and firebrands can waft with the wind. If they strike a house, they may combust, particularly if the structure has a shake-shingle roof. But the more dangerous flame is one that can carry directly from the woods to the walls, which requires a continuous stream of combustibles. Break that continuity, and you break the force of the fire. That interruption can take many forms, but they all fall under the rubric of defensible space. That contrived break makes it possible to stand against the flames and fight back.

Combine this strategy with a sense of habitat, and you have the rough formula for coping with those intermediary landscapes such as ponderosa pine forests that most unsettle the public domain and trouble fire officers. The problem is not one narrowly confined to wildfire but more broadly applies to ecosystem health, or if *health* is too problematic a term, to *ecosystem sustainability*, although that term is no less anthropocentric. Here, concepts of defensible space and sustainable habitat meet. Here, the need to protect human communities while promoting a-human landscapes converge. Call it the search for a defensible habitat or a sustainable space.

These common lands hold a collective theme: their setting is their essence. The issue is not simply putting fire in or taking it out, any more than it is releasing wolves or shooting them. The endeavor will succeed or fail as its context allows. In the past, Mexican gray wolves undoubtedly had roamed, if meagerly, through the lands now claimed by metropolitan Phoenix. Yet no one would consider reintroducing a wolf pack into, say, Arrowhead Mall on the argument that wolves are natural and historically have existed here. However correct the wolves, the existing setting is all wrong: the wolves would not behave as they had in the past. The experiment would end only disastrously for all concerned, not least for the wolves.

So it must be with fire. Simply thrusting torches into forests that have little resemblance to those that existed a century or two previously will not yield fires like those of the past. They likely will either expire in shade or explode into blowups. They can behave as fires should only if the biological setting is right. Here is where the intermix fire scene can segue into a next generation. It has primed the fire community, first, by getting significant fractions of the general population actively engaged with the fire environment and, second, by forcing the debate away from putting fire in or taking it out and into a discussion about what kind of fire environment is suitable. It would be a relatively simple matter to advance this discourse from manipulating the landscape as a means to dampen bad fires into manipulating it to promote good ones.

For the past few decades, both the wild and the urban have grown at the expense of a former rural landscape. The intermix scene has rammed them together, and like matter and antimatter they have detonated. But expand defensible space into a sustainable space, and you can begin to fashion a large buffer zone between the wild and the exurban, and, more important, you can start addressing matters of defensible habitats. The outcome might be a modern avatar of the rural landscape, a different kind of lived-in place, a postmodernist pastoralism, but a place where, unlike the ecological apartheid of today's environmentalism, people, wolves, and fire can find common ground.

paperwork

the literature of forest fires, if there is such a literature

> *Unlike most rangers he was a sensitive, vivid writer, and both his official position and the things in him that saw and felt compelled him to write report after report on the Mann Gulch fire, two of which should be listed in the literature of forest fires, if there is such a listing and such a literature.*
> —Norman Maclean on Ranger Robert Jansson, from *Young Men and Fire*

Why do some national literatures include wildland fire and others do not? Why, in particular, has so little emerged from America's experience with free-burning flame?

The answer to the first is easy. Fire often lies outside literature because in many lands it lies outside experienced nature. The American fire community has become so committed to the belief, by now banal, that fire is natural, necessary, and inevitable that it is worth repeating that fire does not exist spontaneously everywhere on Earth. Indeed, long eons of Earth history passed in which combustibles amassed in geologic stockpiles of oil and coal without succumbing to flame along the way. Although Earth is undeniably a fire planet, fire is neither constant nor predestined.

For humans, however, it is both. Our species monopoly has granted us an ecological agency no other creature possesses. For all our existence, we have held fire. We have never been without it; we have carried it to every place we have gone (which is everywhere, even to the Moon). Perhaps nothing else so expresses our sense of ourselves as the way we apply and withhold flame.

Although fire takes its character from its context, we shape that context and by doing so force flame to obey cultural norms as well as ecological principles, which hauls fire, sputtering and exploding as it were, into the moral universe that people inhabit. There really are good and bad fires.

All of which makes fire an ideal subject for literature—not simply writing that records natural phenomena, but writing that explores the meaning of the world we occupy and the choices we must make about how to live in that world. In fact, we don't need to rely on "natural" fire at all. For literary purposes, unnatural fire offers a cleaner theme because it poses a surer sense of agency. The literature of most nations deals precisely with such fires.

This fact leaves even more mysterious the void within American literature regarding free-burning fire. To see fire in the American experience one has only to look. The West, especially, has superb conditions for natural fire and will burn nicely with or without people. In most of the Northeast, by contrast, people seem to be essential to prepare fuel and kindle spark. In truth, the story in every region is one of people and nature interacting. That's why a literature is possible. And it leaves unanswered the riddle, Why does American literature have so few fire stories?

The answer lies less with nature than with people. The human factor is particularly critical because people must record the flames. They must see them—must choose to see them. They must see some significance, must understand why it matters. And they must write their observations down. Thus, a common paradox: places lean of fire but abundant with litterateurs may contribute to the canon, whereas places rife with lightning fire may warrant less comment about flame than about ticks and mosquitoes. A writer who records fire once likely will record it again. Visitors are more prone to observe new fires, which appear as unusual to them as the local flora or birds, than are residents, for whom the surrounding fires are no more distinctive than the tread of seasons and the bluster of winds. The reverse is also true. Unless fire fits into prevailing fashions and genres, it simply does not enter into formal literature. One cannot scribble accounts of everything; so why fire? Still, one way or another, the observations are there, often as common as beetles.

Consider fire paintings, which mirror the trend with literature. Fires there were before the nineteenth century, but few exist on canvas. A gallery of fire paintings required an enthusiasm for landscapes, particularly wild landscapes; a passion for recording new lands, a documentary rather than painterly spirit; an expanding America that thrust exploring expeditions and far-wandering artists into fire-flushed scenes, such as the Great Plains; and, of course, artists with the temperament to search out flames and understand how they could illuminate image and theme. Thus, almost no fire paintings from major artists exist prior to the 1830s or after the 1890s. Most represent prairie conflagrations, only a few record forest fires; and some artists, such as George Catlin roaming on the American prairies and Thomas Baines perambulating around the British Empire, rendered multiple fire paintings, but others did none at all. Those oils and watercolors testify more to the state of cultural enthusiasms than to the actual status of burning on the land.

Literature is a bit broader because more venues were possible—oral traditions, diaries, newspapers, letters, journals of explorers and travelers, essays, scientific articles, government documents. Most reported on fire as an event of local importance, like a flood or other disaster, or they pondered it as a curiosity or collectible much as Audubon did North American birds. The trickier task was to transmute observation into literature. The natural-history essay has been the traditional means to do so. Oddly, though, until recent years fire has appeared more as a means of change—a plot device, as it were—than as an object of scrutiny itself.

The recent reconsideration of fire has several causes. In part, it reflects the evolving genre, the increasing fascination of an urbanized readership with the exotic energies of nature and the appeal of nature writing as a vehicle for personal essays. In good part, though, it testifies to the apparent waxing of fire on the American scene. In fact, it is vanishing from daily life and vernacular landscapes, and that reality paradoxically makes fire's appearance on select wildlands all the more prominent precisely because it has become uncommon in everyday experience. Wildfire is returning in much the same way and for many of the same reasons as black-footed ferrets and bald eagles. The culture's fascination lies not with fire but with *wild*fire. Here, at least, the public perception is that flame is resurgent, that wildfire looms larger on the land with every season. The flickering flame is becoming seemingly constant. The eyes cannot draw easily away, and they will force the mind and pen to respond.

Nature writing by itself can beg the question of why its examined nature exists as it does. The fact is, America boasts extensive wildland fires because it holds extensive wildlands, and it harbors wildlands because people have chosen to remove themselves from the scene. This decision, backed with political force, is neither obvious nor, paradoxically, natural. It is not natural for humans to vacate a place that they can occupy. It is not natural to remove traces of the human presence. It seems especially odd to hand fire back to Nature after Nature had granted it to human beings as a species monopoly. But a happy accident of geography and history conspired to do just that in America. For a few decades, a relatively vacant landscape existed in the West. The land was emptied of its indigenes and had not yet filled with colonists. The federal government reserved vast patches of that scene from settlement.

By ordaining a permanent public domain, Americans created new habitats for fire, a new human role as fire tender, and a novel literature. In relatively few parts of the world is fire "wild" in the sense of being purely natural. Nearly all fire regimes are a negotiation between cultural and natural forces. Most fire exists on Earth because people choose it; it thrives within agriculture, on field and pasture; it flourishes as a kind of domesticated or captive species, like draft oxen or hunting hawks. Most cultures begin their inquiry into fire with the tamed flame. But not Americans: the originating fire must be natural. All other expressions are in some way a derivation and, most often, a declination.

One would expect that the practice, a century old, of throwing usually young firefighters against the flames would yield a steady outpouring of what that experience has meant. One would think that out of the tens of thousands of firefighters, a robust literature would have bubbled up, if not boiled over. Yet almost nothing exists, and what does exist has emerged recently and often from those not themselves participants. Most of the writers were just that, writers. George Stewart and Norman Maclean were English professors who found in fire a handy plot mechanism or a powerful metaphor. Those who wrote about Yellowstone, South Canyon, or Cerro Grande were journalists by profession. The indigenous literature on fire—efforts by those who have lived on the line—continues, oddly, to stumble. Why?

It may be that big subjects, like big fires, are best attacked indirectly. Those face-to-face with the flames are too close, too busy grasping tools, coughing, or staring dumbly at smoking duff. They think with their hands. They seem unable to stand outside the line and look in, or in their hearts they don't wish to miss the action. This seems peculiar, but then that observation holds true for most war literature as well. America's Civil War produced almost nothing of literary merit, until Stephen Crane, born six years after it ended, wrote *The Red Badge of Courage*. It may be that wildland fire is a subject that one lives or writes about but not both equally well. It may be that the fatigued-to-the-point-of-dumb-exhaustion metaphor of firefight as battlefield casts fire writing into the genre of war literature, with all its liabilities and little of its moral power.

Yet our relationship with fire is far richer than this. Free-burning fire is not simply a "natural" phenomenon occurring on "natural" reserves, hence amoral and thus outside literary review. Those reserves themselves exist because of human choices: what we do or don't do with fire on those reserves is a moral as well as political act. The moral options, without which literature is mere description, continue to lie mostly fallow. Most writers now recognize that what we do to the land is no longer ethically neutral, but they have not tightened the nut of that understanding to the bolt of what we do with fire in the field. The firefight as a brave act in a wrong cause has hardly been explored by serious writers, even though that choice, not the call to arms, lies at the core of American fire. Most firefighters are seasonals. They work for a few months; their years on a crew constitute the "fire season" of their life; to them, fire is comic not tragic—a coming-of-age story not an epic, a fiery *Life on the Mississippi* not an *Iliad* or *Aeneid*.

All this has a practical upshot. Art can interpret and motivate; fire management may need literature more than literature needs fire. After the disastrous 2000 fire season, it should be clear that America has to rethink its relationship to fire from root to leaf. The Joint Fire Science Program may commit tens of millions of dollars to the useful study of fire. Congressional committees may pile up reports by the cord. But genuine reform won't sound throughout the land until fire also finds a literary voice to proclaim it.

The proof of this assertion lies, as does so much of fire literature, with Maclean's masterpiece *Young Men and Fire* (1992). He was wrong about the Mann Gulch fire's impact. The significant blowup was not on August 5, 1949, in the Rocky Mountains but on August 19 in Central Asia, when the Soviet Union exploded its first atomic bomb. That snapped the U.S. federal government to attention and led to serious funding to discover the mechanics of large fires. In truth, the Mann Gulch fire did almost nothing to reform wildland fire protection until forty-three years later when Maclean wrote about it. *Young Men and Fire* became, in turn, the revealed text by which the events of the 1994 South Canyon fire and that epochal fire season overall were interpreted. That inspired gloss led to the reformed federal policy of 1995. Nature had given art something marvelous to contemplate. And then life followed art.

a privileged phenomenon which can explain anything

If all that changes slowly may be explained by life, all that changes quickly is explained by fire. Fire is the ultra-living element. . . . Among all phenomena, it is really the only one to which there can be so definitely attributed the opposing values of good and evil. . . . It can contradict itself; thus it is one of the principles of universal explanation.

—Gaston Bachelard,
The Psychoanalysis of Fire

Five Fires: Race, Catastrophe, and the Shaping of California originated, David Wyatt informs us, with a spark of insight from Robert Hass, who proposed that California be imagined as the story of Four Fires, events "that swept—like a fire—through California to leave a dramatically altered physical and cultural landscape." Such a "scheme" proposed a "way of condensing California's story into a pattern as ordered and economical as a poem, a poem with a central metaphor and a pattern of recurrence." California's cultural history, Wyatt suggests, shares the same cataclysmic rhythms as its fabled landscape, which in a sense it has internalized as an inner fire of unquenchable rage.

The Big Four are the arrival of the Spanish, the Gold Rush, the San Francisco earthquake and fire, and World War II. Wyatt quickly adds a fifth fire: "the fire of race," which is, as it were, the vestal flame from which the others derive. Nothing has so shaped life in California as its "mix of peoples"; this "fifth fire has in fact burned from the beginning of California's recorded history through all the others." The resulting "sequence of events" argues "for a catastrophic history of violent and consuming surprise." The author then adds assorted other real and virtual fires; compounds the elements with discourses about water, illegal immigration, Japanese internment, "twice divorced" women, post–Cold War Los Angeles, Yosemite Valley, the Zoot Suit riots, the Bear

Flag republic, the Rodney King riots; and concludes with a Bruce Springsteen concert at Constitution Hall. Along the way, the informing conceit gets buried like so many metaphorical coals banked in a grey ash of cultural criticism. Yet it reveals what a fire literature shorn from real flame might look like.

Cultural history, as the academy defines that term, and California, as synecdoche for a multicultural American future—these, then, are the means and ends of *Five Fires*. The book is not about historical events but about the ways in which participants and observers, notably writers, remember and record them. Writing gives personal order to the "flash points" of California history and to the smoldering personal rage that is conflagration's pilot flame. "From its beginnings, writing about California has been a deeply metaphorical enterprise, a continuing and contested act of the imagination." California is "a place so dedicated to symbols" that symbols organize life and prompt even riots. The professional historian cannot penetrate to the felt truth of past events. The cultural critic can. A dialogue of voices can replace imposed chronicles. Even riots are "the voice of the unheard." Like a good postmodernist, David Wyatt inserts his own story into the book and in fact makes *Five Fires* into a bid to join the ranks of those creating, listening to, and analyzing California's chorus or cacophony of voices. His merging of personal narrative with public events in the final chapter, "From Watts to South Central," contains the best writing in the book. His analysis of water in *Chinatown* ominously may hold his best sustained metaphor.

Wyatt's chapters build in rough chronology to a thematic climax with the 1965 and 1992 Los Angeles riots, as recorded, respectively, by Luis Rodriguez in *Always Running* and Anna Deavere Smith in *Twilight*. The one recapitulates Wyatt's metaphor of fire; the other epitomizes his celebration of voice as an alternative to history. Wyatt describes Smith's triumph in this way: "Her work is a work of writing and of imagining, and as such, it does its part in changing the world. Insofar as the catastrophes that have shaped California's history are made possible by collective and continuing acts of repression or distanced spectatorship, Smith then joins that company of watchers and listeners who, in the diligence and shapeliness and immediacy of their efforts at memorial, encourages Californians to return upon and understand their past—without, perhaps, repeating it."

Rodriguez concludes that "fire for me has been a constant motif" and so allows Wyatt to bring closure to his own metaphorical structure. The fires out there become internalized. "Throughout his narrative, Rodriguez works toward the recognition that while the condition of his people may be fire, they can learn to channel its energies. . . . Writing his memoir is Rodriguez's primary act of self-kindling. . . . It is precisely in its deployment and mastery of the complex metaphor of fire that *Always Running* demonstrates the capacity of the responsive and responsible individual to meet the violence from without by a salutary and transforming violence from within." Through literature, individuals, like communities, can "educate themselves toward an understanding of their place in history and of the political dimensions of rebellion, thereby externalizing its power as they internalize the fire."

Readers for whom such passages ring with the clangor of insight will welcome this book. For those to whom the judgment seems orthodox and the symbolism scrambled, the pages may resemble a bowl of academically approved granola, grown soggy in a gloss of critical milk. Or to return to David Wyatt's announced informing metaphor, authorial commentary may resemble a kind of rhetorical pyromancy in which a literary shaman stares into the fire and chants the fragments of visions that present themselves. Insights, persons, stories, books, and events arise and vanish like flickering flames.

It didn't have to be this way. Had the author chosen a more (for him) malleable metaphor—the recovered voice, for example—he could have ended with the sound of his own resonances. In one respect, the flawed metaphor doesn't matter. The book is an anthology of literary commentary, and fans of David Wyatt, cultural critics of California, and others will read its messages by an illumination other than firelight. The book suffers from what Bernard DeVoto termed the literary fallacy, the overexuberant claims of literature as a means of cultural analysis. But all writing suffers the limitations of its genre.

The failure lies elsewhere, and it matters because David Wyatt claims that writing is how we organize our world and because he has announced that the fire metaphor is how he will organize his own writing. It is the author who insists, explicitly, that fire does shape the book, all of it, that fire is a suitable means by which to explicate several centuries of California experience; and for this he is accountable. If contested experiences are decided by narrative, voice, and metaphor, if, as Wyatt claims, "the most compelling writing about California reveals its truth through and because of its *style*," then this book might be judged fairly by the author's own style and by the power of his narrative to inform. The failure of an informing metaphor dooms analysis because, in such criticism, the metaphor *is* analysis.

Because it is difficult to look anywhere in California history and not find fire (sooner rather than later), it would seem indecently easy to exploit fire as an informing conceit. California is a virtual fire sermon, its history a scripture of flames in which one might expect to find whatever meaning one chooses. The one requirement is to keep fire in focus. (*Focus*, after all, derives from the Latin word for "hearth." For that matter, *ink* comes from the Greek verb "to burn in," so landscape can become the parchment onto which fire's record is inscribed.) Fire is an ideal conceit for a book about crossed borders, a landscape of disturbance, and contested records. Instead, allusions replace metaphors.

The likely reason is that fire, real fire, is not the source of the metaphor, and that there is no real fire because there is no real place. The four elements that structure the geography of Wyatt's California are not Aristotle's earth, air, water, and fire, but the academy's race, gender, ethnicity, and class. The story occurs between people and within people; California occupies this moral geography because of historical events that brought wildly different peoples together like a compressing piston that finally explodes; but there is no communal fire because the originating fire, the fire on the land, does not exist in the text. The metaphor has no authentic referent. The only voices heard are those that speak in tongues (and not necessarily in tongues of fire). Fire's rage is bottled into the tidy vials of literature.

Literature has replaced real fire as literary criticism has replaced a real literature. *Five Fires* stands for a (blessedly) small body of fire-alluding literature that manages to dispense with fire. The practice is not uncommon in contemporary criticism: reality resides in the mind of the reader, art in the creative act of the critic. What is unusual is to see fire so explicitly marked as the source metaphor. That makes the book fair game for a fire scholar.

The Promethean founder of this odd genre seems to be Gaston Bachelard, whose *Psychoanalysis of Fire* explains why fire is, in fact, an ultimate metaphor and why its use perfectly suits postmodern cultural criticism. Fire is, Bachelard insists, a "privileged phenomenon which can explain anything," one of the "principles of universal explanation," and, most helpfully, a phenomenon that inspires a kind of reverie among those who study it such that it resists rational explanation. This reverie has so progressed that *"fire is no longer a reality for science."* Bachelard thus sanctions a wholly imaginative playing with fire because there exists no objective fire against which to measure our speculations and metaphors. Instead, there is the "poetic mind," which is "purely and simply a syntax of metaphors." The "Imagination" that produces this ecology of symbols "constitutes an autochthonous, autogenous realm." It is self-referential: it exists quite apart from an external world. Within such a conjured world, fire enjoys special standing, a "metaphor of metaphor"—"among the makers of images, the one that is most dialecticized," alone both *"subject and object."* If fire-as-metaphor stands outside an objective reality, so it also stands outside objective criticism because its meaning derives from the web of connectivity the Imagination engenders. In brief, one can do almost anything one wishes with it, and this is pretty much what happens in David Wyatt's *Five Fires*.

This practice may be good enough for an autochthonous, autogenous world of literary criticism. It is not good enough for a world in which real fires incinerate houses, savage forests, and thrive quite apart from anything human beings may do or even think. Although real fire may have vanished as a category of chemistry, it flourishes in forestry and any other subject that must wrestle with a hard Earth and an indifferent biota. Real fire challenges the imposed symbolic fire. Though it is a marvelous index of human presence—a species monopoly, often the indelible signature of habitation—fire transcends humanity. California would still have fires if every human vanished; and because fire is not exclusively human, it cannot be subsumed wholly within a humanized metaphor. Fire transcends the spoken and written word; it obeys a logic of wind, fuel, and oxidizing chemistry beyond allegory. And any effort to exploit it as an extended metaphor must grapple with that naturalistic presence, not merely trace an ecology of texts and the trophic cycle of printed nutrients. Much as the writers of Wyatt's California, full of internal fire, resist narratives imposed on them by others, so also the fires of a California crammed with mountains, high deserts, chaparral and conifers, and Santa Ana winds subvert Wyatt's conceit—resisting, like those writers, with bravura, irony, and guile, but unlike them with a transcendent indifference that makes natural fire at once universal and alien.

This would not matter had Wyatt not placed fire at the center, not made it the vehicle for discourse, not hauled it out of its autogenous closet, and placed historical and physically deconstructing flames into the narrative. Had he left fire strictly as metaphor, it could crackle contentedly in its literary hearth. But when he tries to connect that metaphor with the source fires, those on the land, he must deal with those nontextual flames, and the critic must judge how well he connects them. By using purely symbolic fire to connect everything, he explains nothing.

More important, Wyatt denies his text the full power of metaphoric fire because the experiential fire gets tripped and tangled in the thicket of his imagined flames. For Wyatt, the true fire is the fire in the mind, not the fire in the mountain. Yet those free-burning fires in the hills could have positioned his text outside the self-referential echo chamber of literature, however variously reprivileged and deconstructed. Fire lives along borders and knows no borders. It is both natural and cultural, both fact and symbol, and had Wyatt allowed it to speak (as it were) in its own voice, it could have done everything he asked of it. Instead, fire loses its metaphorical punch as it cascades through his tenuated text, much as water does passing through a sequence of dynamos. Each further abstraction only deposits another image, one with less heat and light and shorn of the power to decompose whole ecosystems.

Rather, we are left with a chasm between real fires on the land and imagined fires in a text, a chasm so wide that it defies even irony to bridge it. Instead of concluding with a concert in which Bruce Springsteen sings from his album *The Ghost of Tom Joad*, positioned around an imagined "campfire," the author might have witnessed the all-too-real fires that regularly ring the Greater Los Angeles area, watched the meltdown of artificial borders, and witnessed an explication on the fallacy of control, a postironic inquiry into the status of humanity as titular and flawed keeper of the planetary flame. Such attention even could have complemented the author's sublimation of Five Fires into the symbolic fire of internalized rage. Wildland fire has become ungovernable; the crisis, at last admitted, has catalyzed a major effort to shift from a strategy of suppression to one of controlled burning in the belief that frequent, small fires will reduce the potential for catastrophic eruptions. The ambition is to restore fire, which is inevitable, while channeling it into beneficent forms. The social analogue is that small, perhaps private rituals of rage, acted out in literature, might prevent social riots.

The confrontation between fire use and fire control is an old debate, one centered, interestingly enough, in California. The effort to reform fire continues in the hills that frame Watts riots and Springsteen concerts. It is a story, one of many, that David Wyatt ignores and so denies his metaphor the historical and factual fuel it needs to propagate. The promised poetry subsides into sullen embers, not readily revived by rhetorical fanning. Ultimately they threaten to reduce the epistemological edifice of literary criticism to lifeless ash.

the old man and the fire

*He no longer dreamed of storms, nor of women, nor of
great occurrences, nor of great fish, nor fights, nor contests
of strength, nor of his wife. He only dreamed of places now
and of the lions on the beach.*
—Ernest Hemingway, *The Old Man and the Sea*

*Even so, there may somewhere be an ending to this story,
although it might take a storyteller's faith to proceed on a
quest to find it and on the way to retain the belief that it
might be both true and fit together dramatically.*
—Norman Maclean, *Young Men and Fire*

No book has influenced wildland fire in America as much as Norman Maclean's
1992 meditation *Young Men and Fire*. No other book comes close—no text,
even Kenneth Davis's classic *Forest Fire: Control and Use* (1959); no specimen of
fiction, including George Stewart's enduring 1948 novel *Fire;* none of a growing
library of personal memoirs, journalistic inquiries, or lurid romances. The 1994
fire season became one of note and of reform largely because Maclean's best-
selling text had primed a national media to the possible meaning of blowup fires
and scorched crews. Maclean demonstrated that the great void in fire management had been cultural: the field had lost contact with its sustaining culture.
Until it found a philosopher-poet, the tragedy at Mann Gulch barely transcended
the subculture of Northern Rockies fire lore. The truth is, fire history, like
fire, takes its character from its context. Mann Gulch had to wait forty-three
years before its flames connected to the right cultural tinder. Since then, it has
become an American icon, its bleached hillside a sacred site, and smokejumpers, a group whose egos were already on steroids, the most celebrated caste in
the fire community. Out of its towering column of text, the book has cast spot
fires across the literary landscape of America. The man who ended his first
book, *A River Runs Through It,* with the line "I am haunted by waters" ended his
life, it seems, haunted by flames.

The book's sheer cultural weight forces one to probe further. What kind of
book is it? Why does fire matter to it? What distortions might it contain? Despite

its insistent obsession that it is a quest for truth, like all art, it does distort: that shaping is what makes it art. Briefly, the text is a kind of meditation. It involves a search for a story, a story being the only vehicle suitable for conveying what Maclean wants to say. One after another, he picks up and tests forms of expression for what they might contribute, and one after another he rejects them as incomplete. He tries history, art, science, even personal memory—all have specific strengths, and all fail to bear, by themselves, the burden he places on them. What he ends with is a story or a story in search of a story. What he has ultimately is the journey itself, the act of literary pilgrimage. That he didn't complete the text is itself a kind of testimonial. The searching and writing, the struggle of mind and heart, were the point, if not the end.

What, then, of fire as a subject? Did the power of the story emerge uniquely out of the flames as Cape lilies effloresce out of ash? Or could something else have served his purposes as well? Reportedly, Maclean had toyed with the Custer massacre before grabbing onto Mann Gulch. That he speaks in universalist terms argues that other events and groups might have served as well. The book is a search for "truth," yes, but one whose revealed facts are made meaningful by explanatory schemas committed to larger purposes. The reasonable explanation is that Mann Gulch did not inspire the search but fit the needs of a quest long percolating through the writer's imagination. The likely spin is that smokejumping—as practiced on August 5, 1949—agreed with an interpretive schema that Maclean already had in his head. A text this rich has layers of meaning, and one is inclined to agree with Robert Frost that a reader is entitled to find whatever in it he wishes. But two schemas seem particularly appropriate.

The first schema is theological, as befits the son of a Presbyterian minister who opened *A River Runs Through It* with the memorable remark that "In our family there was no clear line between religion and fly fishing." Similarly there would be no firebreak between flame and theology. The transition to fire came easily because, like water, it contains an almost fetid symbolism that can bridge events in the woods with a search for ultimate meanings. Maclean's two books thus follow a biblical order: first flood, then flame; first baptism by water followed by a baptism and trial by fire. Religious imagery saturates the text—the jumpers belong to a fraternity "that also has some dim ties seemingly with religion"; the events that unfold at Mann Gulch resemble the "stations of the cross"; their birth from the sky resembles that of another who also died on a hill and who cried, as they did, "My God, my God, why hast thou forsaken me?" And so it goes: the imagery runs through the text so densely that the book nearly approaches allegory.

The core narrative is one of life and death and the meaning they might hold. The Mann Gulch fire serves these ambitions perfectly. The sidecanyon offers a contained world into which the smokejumpers are born from the sky, birthed from a plane complete with an umbilical-cord static line; live briefly if vigorously, convinced of their immortality or at least of their election and oblivious to their ultimate destiny; and then die in a final struggle to attain salvation or fall to the flames. The blowup fire stands for the ineffable "It" at the core of this miniaturized world, a phenomenon that one cannot see directly; that, if experienced, blanks the pilgrim into unconsciousness; that pursues its implacable logic without regard to those in its path. Mann Gulch is, in brief, a world informed by Calvinist theology.

This explains much that is puzzling in Maclean's account, especially his characterization of the smokejumpers (a term that, oddly, he chooses to capitalize) and how they meet their fate. For one, he insists they are an "elite." Certainly in 1949 they were special, but it is difficult to grant elite status when two members had never been on a fire and the foreman, Wagner Dodge, skipped training to repair the district "barn." Elite groups don't behave this way. What Maclean means, but is too tasteful to say directly, is that they stand for the "elect," those souls predestined for salvation or those, in this instance, who believe that being selected as smokejumpers foreordains them to be saved from the final fire. Like good Christians and unlike the drunks who huddle over their bar stools, they are strenuously engaged with the world. The reality is, they will meet a special kind of fire, a blowup that taps a power deeper than garden-variety burns, and will meet their death without an inkling, until the end, that they will die. Their belief in their own election is not enough. To understand why some die and others don't—that is, why three are saved and thirteen lost to the flames—is the core of the quest, along with the still deeper query, How can such a world have meaning?

Most of the jumpers die disbelieving, apparently, that they "owe the world a tragedy," that damnation can happen to them. They become, literally, the "fallen." Two outrun the flames through a crevice in the rocky reef along Mann Gulch's rim. They act on instinct. They see, they do. They have no time for contemplation: only for the belief, instilled early, that they must bolt for the top, never deviating from a single, strenuous path. Walter Rumsey, "a good Methodist," follows his early training and keeps a single thought in his head (like "the voice from inside Mount Sinai"): "The ridge, the ridge." Robert Sallee spots an opening between the rocks and, without looking "either way," takes it. Another jumper, Eldon Diettert, an honors student, evidently studies their crack in the rocks and rejects it, probably with rational cause, and then goes to an intelligent death. But there is no good explanation for why some are saved and others fall. Even Rumsey and Sallee, the survivors, can offer no objective cause. In a world where the prime movers are, like blowups, unknowable, one can appeal only to faith, to the theology that seeks to rationalize those beliefs, and to the compassion of a storyteller to tell them truthfully. Maclean lets the theology splash over the events like water, deferring to readers to decide if it is acceptable.

That still leaves the extraordinary case of Wag Dodge, who evades the relentless flames by setting his own fire and lying down in its ashes. No part of the Mann Gulch saga so obsesses Maclean as this act or so fills the book—its exegesis serves as a climax—and the reason may be that it stands outside the orthodox theological framework, as Dodge does the consuming fires. Dodge seemingly saves himself, and had other crewmen obeyed his urgings to join him, he might have saved them as well. He does it not by foreordained grace, but by his own acts. His daring escape fire suggests that, by courage and cleverness, faith and good works, some kind of personal salvation is possible. There is room for free agency outside predestination. In *A River Runs Through It*, Maclean affirmed that his "father was very sure about certain matters pertaining to the universe. To him, all good things—trout as well as eternal salvation—come by grace and grace comes by art and art does not come easy." Firefighting has its art as surely as fly fishing. So it matters immensely to Maclean and to the thrust of *Young Men and Fire* that Dodge invents the fire as an original act (it is not, as is Rumsey's run, an outcome of early training), that his fire works, and that his deed does not endanger the others. It is, literally, an escape fire: it frees Dodge from joining the others who fall to the world-consuming flames.

Did the escape fire work? Maclean concludes it did. The blowup's wall of flame would have overtaken Dodge had he tried to outrun it. The long chapter with the "mathematicians" at the Forest Service's Missoula fire lab not only offers a scientific analysis of the fire but yields a plot of spread rates in which the movement of the fire, accelerating up the slopes, is measured against the movement of the men. The outcome shows the "inevitability" of the outcome: this was, all conclude, a "race that couldn't be won." Had Dodge not intervened, had he simply clambered up the slopes, the flames would have overrun him as they did the others. The crew's death was thus predestined for reasons outside their control or even comprehension.

Dodge's fire carved out a small sanctuary for himself. Understandably, other crewmen, who knew him poorly, passed him by. The literary among them might have likened his gesture to Ahab's flapping arm while he was lashed to the white whale. It seemed madness. In fact, it seemed odd because it was spontaneous as well as unprecedented. While this depiction works theologically, it succeeds less well as history. Escape fires were a common part of folklore. Most crews had long attacked wildfires through backfires, and a famous incident during the Great Fires of 1910 pivots around a young foreman who saved his crew through a daring burnout in beargrass on Stevens Peak. The truth, however, lies in the theological construction. For Dodge's fire to bear the weight Maclean places on it, it must be novel, sudden, unexpected.

The stickier point is whether Dodge saved himself at the expense of the others. This was a matter of considerable concern bureaucratically and legally because the flip side of agency is liability. If some of the crew died because of Dodge's fire, then tort claims would sprout next to the concrete crosses on the hill. More important, it would deny Maclean the escape he seeks from a world in which young men can die like squirrels amid incomprehensible flames. Dodge's fire is a creative act: he behaves as an artist and thus serves as a model for a professor of English as well as for smokejumpers. The conclusion is that

Dodge's fire worked as he intended and as Maclean wishes and as the book's structure requires.

That leaves the still unfinished business of whether the theological schema is adequate—sufficient in its explanation, adequate in its emotional and psychological catharsis. This is less clear. "Although divine bewilderment addresses its grief to the universe," Maclean notes at the end, "it only cries out to it. It has to find its answer, if at all, in its own final act. It is not to be found among the answers God gave to Job in a whirlwind." Or in a firewhirl. The hint is powerful that such answers lie within confessional beliefs. They are matters of faith, not art. They reside outside the realm of what the written word can convey.

This leads to the second interpretive schema, literature. The story of Mann Gulch is not simply the allegory of a religious drama. As Maclean defines it, the story is also a tragedy, in which compassion assumes the role of faith, which opens the text to another realm of meaning. Indeed, it would be odd if a man who held the William Rainey Harper Chair in English at the University of Chicago did not appeal to literary themes and templates in his storied quest. Theology may be all right for smokejumpers; it is not enough for a storyteller. To follow the "stations of the cross" as a device for re-creating the stages of the tragedy works wonderfully in disentangeling the conflicting tales and endowing them with meaning. A story, a tragedy, however, needs a more literary device to carry the storyteller's own re-creation. "Far back in the impulses to find this story is a storyteller's belief that at times life takes on the shape of art and that the remembered remnants of these moments are largely what we come to mean by life. The short semihumorous comedies we live, our long certain tragedies, and our springtime lyrics and limericks make up most of what we are. They become almost all of what we remember of ourselves." So there are two forms of faith, one grounded in theology, one in art. The story the Gospels tell is different, after all, from the story of how the Gospels got written. Tragedy does not demand theology, but it does require art.

Maclean doesn't inform us what exemplars might inform his story-quest. Perhaps like Dodge he believed that the storytelling was an escape fire to spare him from the implacable flames that left ash and grief and moral bewilderment, and like Dodge he believed his own act unique and unprecedented. He doesn't say, which leaves us free to treat his story as he does the smokejumpers and to suggest an analogue. This is not the same as an explanation: it acts, like all literary criticism, merely as a form of illumination. Not surprisingly for the work of a scholar of Renaissance English, allusions to Shakespeare abound. The surer bond, however, may be to an author unmentioned anywhere in the text yet who as a contemporary appears to have shared common commitments to language, themes, and secular philosophies. The linkage is not only hypothetical and, if confirmed, indirect, but also suggestive, which may be the best we can do and might be enough.

Ernest Hemingway was born three years before Maclean, and he published his celebrated novella *The Old Man and the Sea* three years after Mann Gulch. It is odd to think of them as literary contemporaries because Hemingway blew his brains out a dozen years before Maclean published his first work outside scholarship. But they were, and *Young Men and Fire* echoes many elements of their shared milieu, from its tough, shortened syntax to its fascination with deadly struggle to its admiration for a code hero, who endures in defiance of a larger tragic destiny. In this instance, the protagonist is the author himself. Norman Maclean, commencing his study of Mann Gulch at the age of seventy-four, holds to his quest as Santiago holds to his big fish, even if those choices bring them far beyond the limits of what they know and if, at the end, they return incomplete—Santiago with only the skeleton of a monstrous marlin for his epic ordeal and Maclean with only an unfinished manuscript and scraps of notes. The struggle is itself the story. Maclean's book with equal justification might be titled *The Old Man and the Fire*.

Nothing is easier than to read into a text what one would like to find in it. Postmodern criticism prefers to shatter the text altogether into shards that the critic as creative artist can reassemble as he or she chooses. Yet for all the religious frescoes on its plastered walls, there is a deeper architecture to the narrative because theology by itself leaves too much unexplained, and this particular theology does not grant grace, even to an author, for the struggle alone. The story—Maclean's, not merely the smokejumpers'—cries out for something more than sacramental rite and an assumption of one's own election. Storytell-

ing as art, not theology, storytelling as a means of resolving "moral bewilderment," must appeal to a literary equivalent of sacramental rites. It would be odd
if Maclean did not look to those literary exemplars of endurance in the face of
implacable violence among which he also grew up.

Whether Maclean intended Hemingway as a model is unknown and
probably doubtful, if we even can decide in what sense one can understand a
"model." But they came of age together as writers, and one is probably not far
off that the author of *A River Runs Through It* read "A Big Two-Hearted River"
and other classics by his era's most famous literary fisherman, and that there
exist uncanny, if muted, echoes in his text. For example, Santiago wishes over
and again that "the boy" was with him in the boat to help, and Maclean notes
the pleasure in having a (relatively) young man, Laird Robinson, to assist him,
even if Robinson's Forest Service bosses no more liked the idea than the boy's
parents did his joining Santiago. Maclean's is a softer universe: one that values
compassion, not just toughness, and that allows for faith in its code, not simply
mastery of a craft. It may be that Maclean invented a literary equivalent to
Dodge's escape fire by which to evade the unbearable brutality and emotional
emptiness of a theology that could not explain the jumpers' sacrifice to his
satisfaction. There were ample precedents for Dodge's fire, although he claimed
he knew none of them, and we can believe him. So also we might point out
analogues for Maclean's text and still accept his unquestioned originality. What
endures, finally, is the story of the story.

This is a book about ends. What of its own ending? The death of the
smokejumpers matters keenly: How, exactly, did they die? What did they think
or understand at the end? What meaning can we divine in their parting ges-
tures? Although Maclean has told and retold the story of their final conflagration, he saves the most telling detail for the very end. We can know the jumpers'
"last thoughts and feelings only by indirection," but, he concludes, "we are sure
of the final act of many of them." After they had fallen, blasted by flame, "most
of them had risen again, taken a few steps, and fallen again, this final time like
pilgrims in prayer, facing the top of the hill, which on that slope is nearly east."
"The evidence," Maclean concludes, "is that at the very end beyond thought and
beyond fear and beyond even self-compassion and divine bewilderment there
remains some firm intention to continue doing forever and ever what we last

hoped to do on earth." The religious imagery is unmistakable. There is the hint of a resurrection, for the young men died and rose before their bodies actually succumbed to the burning. Death had not ended their hopes, and, with Maclean to tell their peculiar gospel, their story will continue.

So it is, one might argue, with *Young Men and Fire* and why an interpretation of the book must transcend its overtly religious symbolism. Like those he wrote about, the author died before reaching the ridge. Yet the manuscript went into print, however incomplete in some of its particulars and especially unclear in its opening passages. What mattered was its conclusion, and in that conclusion the book becomes, for Maclean, like those other pilgrims' final convulsive movement, a resurrected gesture of the fallen author. Viewed in this way, the text is not incomplete at all: its story had always been the act of searching, whose end is ultimately unknowable, and the act of telling, which inevitably must be incomplete, at the end of which there is nothing left but tragedy and the response to it of courage and struggle. In the end was the Word, and to understand this marvelous book one must look to literature as much as to theology.

Perhaps, when his own moment came, Norman Maclean remained haunted still by waters. His wife's death by suffocation, which concludes the text, is likened to a drowning. Perhaps, though, he heard the roar of a blowup fire, the "It" at the core of the universe pressing its heat on his back as he sought salvation up the hill of his long, unfinished quest. Or perhaps he was listening to lions on the beach.

green
skies
of
montana

If the literature of forest fires is sparse, fire films are even thinner on the
ground. (Atrocities such as *Firestorm* hardly deserve status as a cartoon. It's to
wildland fire what *Batman* is to law enforcement.) Why are the pickings so
poor?

One reason, perhaps the most fundamental, is the difference between
action and drama. Fire control offers plenty of physical action, yet it remains
strangely silent regarding moral drama. Rather, wildland firefighting appears as
one of America's great contact sports; hotshotting belongs with stock-car racing
as a kind of public entertainment. The books that spark from time to time read
like juvenile sports stories, about ardent youths making the team or coming of
age during the big game. The fire community tends to *do*, not to write. The
possibilities are rich for a literature of action and adventure. In particular, it is
easy to massage the firefight into the genre of the war story.

For films, there is a second difficulty, which involves directing fire on the
set. (In comparison with wildfire, even actors such as Marlon Brando seem
positively domesticated.) Probably computer animation can overcome this difficulty. But that still leaves unresolved the question of a moral void. Where is the
moral agency? Where is the conflict? Where the choice? One common solution
is to have a human agent start the fire, which then serves as a proxy for human
malevolence. But what if, as in wildland fire, nature kindles the originating
blaze? What justifies throwing people out of airplanes to fight lightning-kindled
blazes in places so remote that they are beyond normal human life? The fires
threaten people only because firefighters attack them. Is this Ahab's pursuit of
a white whale or a simple excuse for adrenaline? The fire becomes a challenge,
not a choice. A smoke report sounds a call to arms and a search for the

courage to face the flames—to attack the fire, any fire, any place, any time. Most routine war dramas demand no more: the value of the fight is both implicit and unequivocal. Great war literature goes far beyond and plumbs the possibility of choice, particularly when every choice is awful.

All this aptly characterizes what is probably our best fire film, *Red Skies of Montana,* a 1952 release, likely inspired by George Stewart's 1948 novel *Fire* and by the 1949 Mann Gulch fire. Its understood subtext, however, is the Korean War. Aside from its clunky special effects, *Red Skies* probably does as much as one can with the genre, and it is right to shift the source of the moral action, the conflict, within the protagonist. The story opens and ends boldly—actually with a near repetition of a blowup fire—but creaks through the flabby middle as the physical action slows, unable to find an internal drama as strenuous as fireline duty, and we must watch smokejumper foreman Cliff Mason struggle to overcome his amnesia and self-doubt—not the best material for a *motion* picture, which after all thrives on movement. What redeems the movie are some surprisingly strong acting; several striking sets, notably the burn on Bugle Peak like a World War I no-man's land and the parachute loft, where chutes hang limply like ghosts; a haunting musical score; and the simple momentum of a firefight.

Not much changed over the next forty years until Norman Maclean turned the young smokejumpers who died at Mann Gulch into existential heroes, engaged in an unequal fight with the transcendent "It" at the core of human existence. The oxygen-sucking blowup became a physical metaphor for the unknowable meaning at the core of the universe. The moral story thus turned inward. It's not obvious, however, how one might film such a meditation. If someone does, the outcome likely will resemble *Red Skies,* but with better graphics and a two-income family.

Yet an alternative story does exist, and it derives from reexamining our relationship to fire and indeed to nature. When *Red Skies* was released, America was engaged in a cold war on fire; all flame could be imagined as enemy fire, as hostile fire. The only administrative requirement was to attack those flames as quickly as possible, to apply a doctrine of "force enough, fast enough." But that perception no longer holds, not ecologically, not ethically. What we do or don't do with nature is no longer considered neutral: nature is not simply a backdrop or a handy crank to turn the plot. What we do with fire has consequences far greater than the individual choice to jump or not to jump. Attacking fires in the backcountry may be a mistake—and is now regarded as a regrettable error of national judgment. If the firefight is a moral equivalent of war, it equates with Vietnam rather than with World War II. Smokejumpers have thus become valiant workers in a deeply flawed cause. They are the fire equivalent of big-tree loggers, market hunters, and cowboys hired out to cattle barons. The western landscape may be the worst for their labors.

All this offers the prospect for a drama more nuanced than the simple and battle-fatigued firefight-as-battlefield metaphor. Today, wildland fire is awash with choices, all mired in complexity, none of them answerable by a simple clarion call to suit up and jump. No one yet has found a way to tell this narrative in literature or film. When, in the climax to *Red Skies*, Cliff Mason holds his crew in their foxholes by fist and force of personality, he is replaying the central scene of American fire history, that ineffable moment when Ranger Ed Pulaski held his crew at gunpoint in a mineshaft while the firestorms of the Big Blowup of 1910 raged about them. That event is a magnificent piece of Americana, but it barely whittles at the immense complexity of a relationship to fire that compels us to choose when and where and how to apply and withhold fire and what to do to the land to make it receptive to whatever fire regime we determine belongs. Such qualms never crease the brows of the dispatchers and fire officers and smokejumpers of *Red Skies*. But they should.

Still, if firefight-as-battlefield is a tired metaphor, sucking intellectual life from stories as a blowup fire does oxygen, there is none of equal stature to challenge it. There should be. What America needs is a literary equivalent to the scientific recentering of fire within biology. It needs to shed the old genres and formulas entirely, particularly one as hand-me-down and hackneyed as the firefight. The moral action lies not in the fighting but in the choosing. Contemporary environmentalism suggests a way this might be done. We need someone who can turn Montana's red skies green.

debriefing

why
i
do
it

The only mystery is why I wouldn't. I had long lived two lives, one in the academy, the other on the North Rim. It finally soaked into my dense-barked skull that what I ought to do is apply the scholarship I'd learned in the one place to the subjects of the other, particularly fire, which is what most gripped me. I'd smokechase with a pencil rather than a pulaski, with a firepack laden with notecards, archive passes, and reports instead of flagging tape, C-rations, and files. I've never stopped.

In October 2001, the board of directors of the Forest History Society convened for its biannual meeting, this time in Victoria, British Columbia. We met in a conference room of the Royal B.C. Museum—an edifice, like all modern buildings, designed to prevent fire or, if one starts, to evacuate inhabitants briskly. There were emergency exits, emergency lights, emergency alarms, and sprinklers. Within, we learned about the year's awards. The Theodore Blegen Award went to Andy Fisher for an article that detailed the history of the Southwest Forest Fire Fighters program. The John Collier Journalism Award went to Sherrie Devlin of the *Missoulian* for her series on the fabled Big Blowup that savaged the Northern Rockies in 1910. The Charles Weyerhaeuser Book Award was captured by John McNeill, whose *Something New under the Sun* tracks the environmental history of our fossil fuel–driven globe. The business meeting having concluded, we set out, by diesel-combustion bus, on a field trip into Vancouver Island.

The engine on the bus caught fire. The driver tried to ignore it, but smoke seeped into the rear seating, and opening the emergency vents would not clear it. Eventually he had to pull over at a wide swath of the highway. While he called for a replacement vehicle, we looked around. We were maybe two hundred yards from a B.C. Forest Service fire cache. We walked to it, and the duty foreman gave us an impromptu tour and lecture about fire protection in the south half of the island. Back on the road, we pulled off at the Goldstream Park, a picnic area along a salmon run. The place had been overrun by the massive 1938 fire complex. Everything we saw dates since that catastrophe. Our hosts served lunch farther along, at the Cowichan Lake Research Station. Across the waters, we could watch slash burning on the hills. We visited the nursery and an extant patch of dark and dripping old growth, which originated in the ashes of drought-propelled fires during the seventeenth century. Everywhere we cared to look, there was evidence of combustion and, once outside of museums and hotels, of honest flame.

Why study fire? Because it's there. Because it's fascinating. Because it matters. Because it's fun. Because if we can no longer get excited about fire, then we might as well resign our species membership in the Great Chain of Being.

acknowledgments

"Hominid Hearth" was originally written for *Handbook of Fire Management in Sub-Sahara Africa,* edited by Neels deRonde and J. G. Goldammer (forthcoming), and is reprinted with permission. "Old Fire, New Fire" originally appeared in *ISLE* 6, no. 2 (summer 1999), and is reprinted with permission. "Strange Fire: The European Encounter with Fire" was originally published in *20th Tall Timbers Fire Ecology Conference Proceedings* (Tallahassee: Tall Timbers Research Station, 1998) and is reprinted with permission. Another version appeared in *Ecology and Empire,* edited by Tom Griffiths and Libby Robin (Seattle: University of Washington Press, 1997). A version of "America's War on Fire" appeared under the title "Metaphoric Meltdown; or, Not Back in the Saddle Again," in *Wildfire Magazine* (November 1997) and is reprinted with permission. A short version of "When the Mountains Roared, Again" was published in *Under Fire: The West Is Burning,* edited by Tom Fenske (Billings, Mont.: Tom Fenske, 2001), and is reprinted with permission. "Burning Deserts" was written for *Arizona Highways* but never published and appears now with permission. An early version of "Compassing About with Sparks," under the title "The Perils of Prescribed Burning: A Reconsideration," appeared in *Natural Resources Journal* 41, no. 1 (winter 2001), and is reprinted with permission. A slightly different version of "Smokechasing: The Search for a Usable Place" was first published in *Environmental History* 6, no. 4 (October 2001). This essay was written for the Distinguished Lectureship in Forest and Conservation History sponsored by the Forest History Society, the Duke University Department of History and Nicholas School of the Environment, and the North Carolina Humanities Council. A much longer version of "An Exchange for All Things?" was published in *Australian Geographical Studies* (March 2001) and is reprinted with permission of the Institute of Australian Geographers. A version of "Green Skies of Montana" first appeared in *Forest History Today* (spring 2000) and is reprinted here with permission of the Forest History Society. "Buy a Book, Save the Planet" appeared as "A Modest Proposal" in *Authors Guild Bulletin* and is reprinted with permission.

about
the
author

Stephen J. Pyne grew up in Phoenix, Arizona, took his undergraduate degree at Stanford University, and acquired a Ph.D. from the University of Texas at Austin. He has been a professor at Arizona State University since 1985, ten of those years at ASU West. He is presently in the Human Dimensions in Biology Program of the School of Life Sciences at ASU Main. For fifteen seasons, he worked on the fire crew at the North Rim of Grand Canyon National Park—twelve as foreman—and later spent three summers writing fire plans at other parks. His research has focused on the history of exploration, the Grand Canyon, Antarctica, and, especially, the global history and management of fire, about which he has written eleven books. He also contributed a foreword to the University of Arizona Press edition of Clarence Dutton's *Tertiary History of the Grand Cañon District* (2001).

Also by Stephen Pyne:

Fire: A Brief History

Year of the Fires: The Story of the Great Fires of 1910

How the Canyon Became Grand: A Short History

America's Fires: Management on Wildlands and Forests

Vestal Fire: An Environmental History, Told through Fire,

    of Europe and Europe's Encounter with the World

World Fire: The Culture of Fire on Earth

Burning Bush: A Fire History of Australia

Fire on the Rim: A Firefighter's Season at the Grand Canyon

The Ice: A Journey to Antarctica

Introduction to Wildland Fire, two editions

Fire in America: A Cultural History of Wildland and Rural Fire

Grove Karl Gilbert: A Great Engine of Research